SOTHIS PRESS

Sothis Press aims to completely reimagine the public domain book with an emphasis on universal access to captivatingly designed print editions. We believe that technology will enable every single book ever published to be perpetually accessible in beautiful print form. Sothis is beauty in ubiquity.

www.sothis.press
1 Waterhouse Square, London, ECIN 2ST

Navigation and Nautical Astronomy

With Special Table, Diagram, and Rules
Adapted for Navigating Iron Ships

W. T. Read

SOTHIS PRESS

ISBN 978-1-0365-1037-4

This edition is published by
Sothis Press © 2023. The company
is registered in England and Wales.
The Company Number is 14900389.

TO

CAPTAIN J. F. TRIVETT, LIEUT. R.N.R.,

LATE OF THE

STAFF DEPARTMENT, BOARD OF TRADE,

FIRST COMMANDER OF THE "WORCESTER,"

WITH WHOM

SEVENTEEN YEARS SINCE, IT WAS THE AUTHOR'S PRIVILEGE

TO LABOUR IN ESTABLISHING THE SCHOOL,

AND WHO,

AS A MEMBER OF THE COMMITTEE OF MANAGEMENT,

HAS EVER SHOWN AN ACTIVE INTEREST IN THE EDUCATION OF THE CADETS,

THIS VOLUME

IS RESPECTFULLY DEDICATED.

NAVIGATION

AND

NAUTICAL ASTRONOMY,

WITH SPECIAL TABLE, DIAGRAM, AND RULES

ADAPTED FOR NAVIGATING IRON SHIPS.

BY

REV. W. T. READ, M.A.,

HEAD MASTER AND CHAPLAIN,

THAMES NAUTICAL TRAINING COLLEGE, H.M.S. "WORCESTER."

LONDON:

ELLIOT STOCK, 61 & 62, PATERNOSTER ROW.

1879.

Price Ten Shillings and Sixpence.

S. BLEACH, Printer,
23, Little Bell Alley,
Moorgate Street,
E.C.

PREFACE.

It has been the Author's object in the following pages to produce a Course of Navigation and Nautical Astronomy which shall enable the student intelligently to master the principles upon which the various methods of finding a ship's position at sea are founded, while he may endeavour to secure accuracy of computation.

A moderate knowledge of Mathematics is necessary to a full understanding of the Theoretical teaching here presented, but it is hoped that this work will prove of value, also, to those whose mathematical knowledge is but limited, and that the manner in which the various problems are arranged will bring it within the comprehension of seafaring men generally.

The following course is, in reality, a representation of that which has been uniformly adopted in the training of Mercantile Marine Cadets on board H.M.S. "Worcester" (Thames Nautical Training College), over the Mathematical instruction of whom, it has been the Author's privilege, since its institution, to preside, and he is induced to offer it as the result of many years' experience in the teaching of Theoretical Navigation and Astronomy.

The introduction of iron vessels, in later years, and their continued multiplication, both in the Navy and Merchant Service, demand an intelligent acquaintance at the Navigator's hands, with the principles of Deviation of the Compass, and all Courses corrected, practically, at sea, are treated from a similar Table of Deviation to that presented in this work.

It has been thought advisable, therefore, that the student should accustom himself to the Correction of Courses from a Table.

The Author has had many opportunities of observing that, though correcting courses in this manner may seem simple, it requires considerable practice, upon the part of young beginners especially, to ensure safe results, both in working the ordinary "Day's Work," as also in the converse process of finding the Compass from the True Course.

All the examples in the Sailings, therefore, in which Local Deviation is involved, have been worked from the table given at page 12.

It should be noted that in correcting a compass course, the *Deviation* is the *first* correction to be applied, being obtained from the given position of the ship's head.

The "Methods" of Nautical Astronomy are arranged in the order of Latitude, Longitude, Time and Azimuth; but, if it be preferred to study, *e.g.*, a method in Longitude before another in Latitude, there is no reason why intervening pages should not be temporarily missed, as there is no necessary connection between some of the problems.

The Tables employed in the solutions are used in most nautical works, but those of Riddle are especially recommended, and with them the work has been compiled.

It is hoped that these pages will be found to contain all that should be required at the hands of an educated and intelligent ship-master.

The time will, doubtless, sooner or later arrive, when the "Rule of Thumb" methods, so popular up to the present, must be abandoned, and those who hold responsible positions as Officers in the Mercantile Marine Service, shall be required, at the Board of Trade examinations, to show that they understand the *reasons* upon which those rules are founded.

To provide such officers, has been its promoters' object in the institution of the "Worcester" Nautical College which has spent seventeen years of useful existence in the welfare of the Mercantile Marine Service of the Country.

That Institution has sought to follow up, for the most part, the course of education given in the old Nautical School of Greenwich Hospital, where, under that eminent teacher and Mathematician, the late Mr. John Riddle, F.R.A.S., and his father previously, many of the Navigating Officers of the Royal Navy, and able Commanders of the Merchant Service were trained for their profession.

The Author trusts, in conclusion, that the present work may be found serviceable both to Navigating Officers, and to pupils in course of training in Navigation Schools generally.

The Examples at the conclusion of this Work are entirely *original*, and were prepared with a view to the present publication.

CONTENTS.

PART I.—NAVIGATION.

H Conway

CORRIGENDA.

Page 8, line 13, for "Semi-circular," read "*Semi-circle.*"

" 8, line 19, for "mininum," read "*minimum.*"

" 13, Ex. 1, for SW $\frac{1}{4}$ W, read *SW* $\frac{1}{2}$ *W.*

" 24, Ex. 14, for S, read *Southerly.*

" 26, Ex. 6, Lat 38° 7′ 6″ N.

" " " Co S 51° 38′ W : dist 56·23.

" 28, Ex. 6, Lat 54° 18′.

" 31, Ex. 7, Comp co N 33° 19′ W.

" 31, Ex. 10, Dist 408·2.

Longitude.

Time.

Latitude and Longitude.

Lesser Computations.

Variation of the Compass.

PART III.—INVESTIGATIONS OF RULES.

PART I.

NAVIGATION.

NAVIGATION.

Navigation is the science which enables the mariner to conduct his vessel over the surface of the ocean from port to port, and also to calculate his position at sea when occasion requires.

This science is divided into Geo and Cœlo-navigation, the former part is generally known as the dead reckoning, the latter embraces a knowledge of the methods of determining latitude and longitude (or position of a ship), by means of observations of the heavenly bodies, and is called Nautical Astronomy.

Sufficient evidence has been adduced to establish belief in the rotundity of our earth. The following arguments are generally considered conclusive as demonstrative of the theory, viz :—

I. In looking for the appearance of a vessel on the horizon, we find that the higher parts of the mast are first visible, then, gradually, the hull of the ship comes into view : or, again, in receding, the hull disappears first and then the higher portions. The body on which such a phenomenon takes place must necessarily be of a spherical form.

II. Navigators have been known to compass the world by sailing in the same direction, and have returned to the place whence they set out.

III. In eclipses of the moon, (caused by the projection of the earth's shadow on her disc), the figure presented by the shadow is invariably round, which could only be projected by a spherical body.

The earth is, in fact, an oblate spheroid, (*i.e.*, a spherical body whose transverse diameters are unequal), the polar diameter being 26 miles less than the equatorial, (the former 7899, the latter 7925 statute miles). Our earth is the third planet of the solar system, whose elliptical path or orbit round the sun, is traversed in 365¼ days, and it revolves on its axis once in 24 hours.

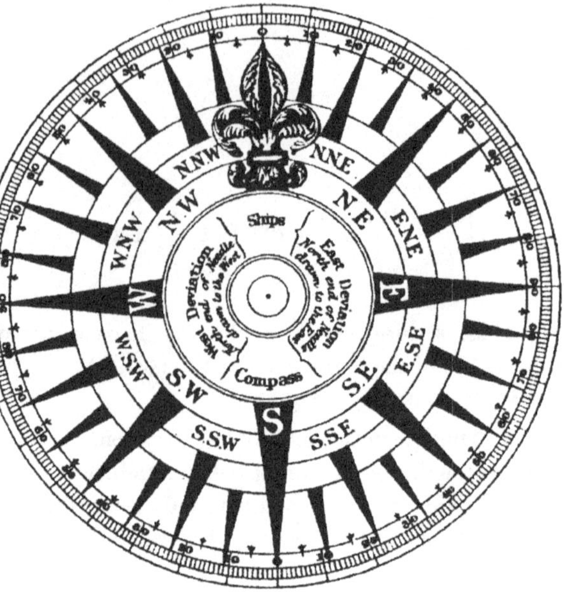

The subjoined tables give the points and quarter points of the Compass and their values in degrees :—

N—E	N—W	S—E	S—W	Pts.	° ′
North	North	South	South		
N ¼ E	N ¼ W	S ¼ E	S ¼ W	¼	2.48
N ½ E	N ½ W	S ½ E	S ½ W	½	5.37
N ¾ E	N ¾ W	S ¾ E	S ¾ W	¾	8.26
N b E	N b W	S b E	S b W	1	11.15
N b E ¼ E	N b W ¼ W	S b E ¼ E	S b W ¼ W	1¼	14. 3
N b E ½ E	N b W ½ W	S b E ½ E	S b W ½ W	1½	16.52
N b E ¾ E	N b W ¾ W	S b E ¾ E	S b W ¾ W	1¾	19.41
NNE	NNW	SSE	SSW	2	22.30
NNE ¼ E	NNW ¼ W	SSE ¼ E	SSW ¼ W	2¼	25.18
NNE ½ E	NNW ½ W	SSE ½ E	SSW ½ W	2½	28. 7
NNE ¾ E	NNW ¾ W	SSE ¾ E	SSW ¾ W	2¾	30.56
NE b N	NW b N	SE b S	S W b S	3	33.45
NE ¾ N	NW ¾ N	SE ¾ S	SW ¾ S	3¼	36.33
NE ½ N	NW ½ N	SE ½ S	SW ½ S	3½	39.22
NE ¼ N	NW ¼ N	SE ¼ S	SW ¼ S	3¾	42.11
NE	NW	SE	SW	4	45. 0
NE ¼ E	NW ¼ W	SE ¼ E	SW ¼ W	4¼	47.48
NE ½ E	NW ½ W	SE ½ E	SW ½ W	4½	50.37
NE ¾ E	NW ¾ W	SE ¾ E	SW ¾ W	4¾	53.26
NE b E	NW b W	SE b E	SW b W	5	56.15
NE b E ¼ E	NW b W ¼ W	SE b E ¼ E	SW b W ¼ W	5¼	59. 3
NE b E ½ E	NW b W ½ W	SE b E ½ E	SW b W ½ W	5½	61.52
NE b E ¾ E	NW b W ¾ W	SE b E ¾ E	SW b W ¾ W	5¾	64.41
ENE	WNW	ESE	WSW	6	67.30
E b N ¾ N	W b N ¾ N	E b S ¾ S	W b S ¾ S	6¼	70.18
E b N ½ N	W b N ½ N	E b S ½ S	W b S ½ S	6½	73. 7
E b N ¼ N	W b N ¼ N	E b S ¼ S	W b S ¼ S	6¾	75.56
E b N	W b N	E b S	W b S	7	78.45
E ¾ N	W ¾ N	E ¾ S	W ¾ S	7¼	81.33
E ½ N	W ½ N	E ½ S	W ½ S	7½	84.22
E ¼ N	W ¼ N	E ¼ S	W ¼ S	7¾	87.11
East	West	East	West	8	90. 0

THE MARINER'S COMPASS.

First among the instruments which enable the mariner to make his way in safety from shore to shore, is the compass. Much might be said concerning this most useful of all nautical appliances, but, as our space is limited the facts connected with it must be presented in a condensed form.

It may be sufficient, however, to notice that, in consequence of the yet imperfectly developed state of the science of magnetism there are phenomena connected with the compass which are still unexplained. Experience, of late years especially, has shown that no detail connected with navigation demands more study and watchfulness on the part of the navigator than this simple instrument, and the conditions by which it is affected in iron ships.

A skilful seaman should make it his aim to perfect his knowledge of that branch of modern science which treats of the varied action of iron upon the suspended magnetized needle, and to become fully acquainted with the manner in which a compass behaves in a vessel constructed, more or less (as vessels now are), of iron.

The compass card is a thin, circular sheet, generally of cardboard, supported so as to turn freely and readily about its centre. The bearing, or point on which it rests is generally a small plate of agate (sometimes a ruby), let into the card with a conical hole at the centre resting upon a fine needle-point of hard steel; by this arrangement, any friction is, as far as possible, diminished. The edge of the card is divided into thirty-two equal parts called **rhumbs**, or points of the compass, and also into 360 equal portions, called **degrees**.

If a magnetized bar of steel be carefully poised on its centre of gravity and be allowed to assume whatever horizontal position it may choose, we shall find that one end is invariably directed to a particular point of the horizon, approximating in position to the true north. This point is known as the **Magnetic North**, and for the sake of convenience, the extremity of the balanced needle directed thereto is marked as the north end. Now such a balanced needle or magnet, (or, in the best instruments, a series of parallel magnets), is placed beneath the compass card, so that the north end shall coincide exactly with the point marked north on the card, and the south end with the south. The card so balanced is now placed in a hemispherical bowl of thick copper which is called the compass box, it having been found that the presence of a mass of that metal serves to damp the vibrations of a freely suspended needle, without interfering in any way with its freedom of motion. The compass box is suspended in **gimbals**, which consist of two concentric copper rings, the outer one resting on a horizontal axis secured at opposite ends to the compass case, by means of screws: the inner ring has its axis at right angles to that of the other and is secured to the outer ring in that direction: thus supported, the compass remains horizontal and unaffected by the rolling or pitching of the ship. There is reason to believe that this property of the compass needle, viz., its pointing to one particular part of the horizon, was known to the Chinese 2000 years before the Christian era. The angle at the centre of the compass between the true and magnetic north is called the **Variation of the Compass**, which is subject to slow and gradual change in long periods of time. At present, the north end of the needle is directed to about two points to the west of the true north. This change in the variation is accounted for by the theory that the magnetic pole makes a slow revolution round the pole of the earth, followed, in its course, by the end of the compass needle.

MAGNETISM OF IRON SHIPS.

This is a subject of such importance to the seaman, and one which recent scientific investigation has linked so closely with the study of Navigation, that a few words must here be added in reference to it.

We have said that a magnetized bar, if left to traverse horizontally about its centre, will settle itself in the direction of the magnetic north. Now let a similar bar be poised so as to travel vertically, and let it be held in the plane of the meridian, *i.e.*, in a north and south direction. It will at once dip downwards in this latitude, until it makes an angle of 70° with the horizontal line. This angle is called the **Magnetic Dip**, and the direction of the bar when at rest, is the **Line of the Earth's Magnetic Force** at that particular place. The lower extremity will have been attracted by the magnetism of the earth, as one magnet attracts another ; and in this hemisphere, that extremity will be marked as the North Pole of the magnetic needle, though, upon the principle that opposite poles of magnets attract one another, it is really the south end of the magnet attracted downwards by the northern magnetism of the earth.

In order to facilitate remembering the polarities of a magnet, North magnetism is called **Red** (from the **R** in north), and South magnetism is termed **Blue**, (the letter **U** occuring in South). Frequently the extremities of a magnet are painted with the colours respectively.

Imagine such a needle to be carried round the earth from pole to pole. There will be two almost diametrically opposite points, each in a high latitude, and approximating to the true poles of the earth, where the needle would stand vertically. These points are called the **Magnetic Poles**.

There will also be an irregular curve, crossing and recrossing the Earth's Equator, where the needle will remain horizontal: this curve is called the **Magnetic Equator**.

If then, a bar of iron be held in the direction of the dipping needle at any place, and be hammered smartly for some time, it will be found to have acquired a considerable amount of magnetism, and will possess all the properties common to the magnetic needle. The process by which it thus receives magnetism from the earth, is called Magnetic Induction.

This fact throws light at once upon the method by which an iron vessel becomes, whilst building, one vast magnet.

Subjected as she is to the hammering necessary in her construction, she receives a quantity of magnetism, a part of which never leaves her, and has therefore been termed Permanent Magnetism: that part which disappears after some interval, has been called Sub-Permanent Magnetism.

Every iron vessel, when fitted for sea, is furnished with an Azimuth Compass, *i.e.*, a compass of superior construction, and, moreover, fitted with a sight vane and reading prism so as to be used for observing amplitudes of the sun or other heavenly bodies in determining the deviation. This compass is generally raised at a considerable height above the deck, so as to be beyond the influence of temporary derangement by passing magnetic attraction, and is the compass with which the Binnacle and all other instruments of the kind are compared. It remains unaffected by anything but the variation and the deviation natural to the ship as she is fitted in readiness for service and may therefore be safely trusted. This Standard Compass, as it is called, having been placed in its permanent position, the vessel is taken into still water—*i.e.*, beyond the reach of tidal influence—and the process of Swinging Ship to determine the deviation on every point, (generally), of the Compass, is commenced, and is accomplished as follows,—

I. By Cross or Reciprocal Bearings.

A standard compass is taken to the shore and placed in such a position as to be free from the influence of iron or magnetic attraction, and where the ship's compass may be seen. The vessel's head is then brought round, by means of warps, to each point of the compass successively, and simultaneously the bearings of the shore compass from that of the ship, and of the ship's compass from that on shore are taken for every point: the difference between the readings being the deviation for that position of the ship's head: east if the bearing of the shore compass be to the right of that on board, and west, if to the left.

Or II. The standard compass may be taken on shore to a spot whence the part of the ship where her compass is fixed, and a distant object whose bearing is known, shall be in one and the same straight line. The bearing of that object from that point will be the same as its correct magnetic bearing from the ship.

Or III. From an observed azimuth of the sun, by methods hereafter explained, the error of the compass may be deduced: then, deducting the variation from a chart in which the variation is laid down for every place on the globe, the remaining portion of the error of the compass, will be its deviation.

The deviation being thus obtained for each position of the ship's head, it is tabulated and kept for the future correction of courses; or it is laid down upon what is known as Napier's Diagram, from which, at sight, the true course being known, the steering course may be determined, and vice versa.

The deviation in an iron vessel has been found to be composed of two principal parts, caused by magnetic forces at right angles to one another. These parts are known as the semi-circular and the quadrantal deviation, the effect of the former being largely represented in the Napier's Diagram. As an iron vessel is swung round the compass, the principal part of her deviation is easterly in one semi-circular and westerly in the opposite one, the line of neutrality coinciding very nearly with the direction in which her head was while in building, and the deviation reaching to a maximum when she is at right angles to this direction. When therefore, an iron ship is steering on a course corresponding to her position while building, she may be said to be in the line of her own magnetic force, and the effect of her magnetism on the standard compass will be a mininum. This semi-circular part of the deviation changes when the ship proceeds to a different latitude, and the compensation which is generally made by magnets round the compass, is reversed when she enters the opposite hemisphere. The semi-circular deviation is caused chiefly by the action of vertical iron, such as stanchions, an iron stern-post, &c.—for as we have already intimated vertical iron receives magnetism from the earth by induction, the lower end of such iron being, in the northern hemisphere, a north pole. Iron also which lies coincident with, or parallel to the fore and aft line, or at right angles to it, produces the same effect upon the compass.

Quadrantal Deviation is so called from being easterly and westerly in alternate quadrants, as the ship's head swings round a complete circle: its neutral points coincide very nearly with the cardinal points of the compass, and it is caused chiefly by the action of iron lying diagonally,—*e.g.*, from one bow to the opposite stern quarter. When once compensated for, it is never liable to change, and is unaffected by the change of place.

These two principal parts of the deviation are represented by four co-efficients B, C, D, E, the first two comprising the effect of the semi-circular, and the others the Quadrantal. Their values may be easily remembered from the following expressions

$$B = \frac{\text{Dev. at E} + \text{Dev. at W}}{2}, \text{ changing the sign of latter.}$$

$$C = \frac{\text{Dev. at N} + \text{Dev. at S}}{2} \quad \text{''} \quad \text{''} \quad \text{''}$$

$$D = \frac{\text{Dev. at NE} + \text{Dev. at SW}}{2} \quad \text{''} \quad \text{''} \quad \text{''}$$

or

$$D = \frac{\text{Dev. at SE} + \text{Dev. at NW}}{2} \text{ changing the sign of latter.}$$

$$E = \frac{\text{Dev. at N} + S + E + W}{4}, \text{ reversing the last two signs.}$$

a small portion of the entire deviation, called A remains, but is generally considered insignificant in value,

$$A = \frac{N + S + E + W}{4}$$

It should be understood that easterly deviations are reckoned as +, and westerly as — : and the Algebraical sum is always taken in computing the above co-efficients.

As an iron vessel heels over, by the influence of wind, or otherwise, her compass is liable to another considerable error, called the **Heeling Error.**

In heeling she approaches the line of the earth's magnetic force, consequently there are forces now developed by her beams becoming inclined to this line, which, when she lay on an even keel were dormant. Two or three remarks may be of use, in reference to this error.

(1.) In Northern latitudes, as the ship heels over, the compass needle is drawn to windward; this is a fact which should never be missed.

(2.) Whenever the ship's head is E or W, by the disturbed compass, the heeling error vanishes, and is greatest when it is N or S.

(3.) The amount of heeling error is determined by observing the vibrations of a dipping needle, first on board, then on shore, in a given time. If T be the time say of 10 vibrations on shore, and T^1 the time of the same on board, we have,

$\frac{T^2}{T_1^2} = \frac{Z^1}{Z}$ which is the relation that the mean value of the vertical magnetic force at the place of the compass bears to the earth's magnetic force.

The following useful rules are invaluable with reference to the Standard Compass.*

1. The Standard Compass not to be within half the breadth of the ship from the rudder-head and stern post, or iron cased screw well: not to be nearer an iron deck, or iron deck beams than five feet.

2. In ships built nearly head north, to be as far forward as the requirements of the ship will permit. In ships built near head south, to be as far aft as circumstances may permit. When built head east or west, not to be near either extremity.

3. No masses of iron, as boilers, stanchions, &c,, to be placed near the compass, or within 50° of the vertical line through the centre, the angle being drawn from the centre of the compass to that of the mass in question.

*See "Elementary Manual for Deviation of Compass in Iron Ships," by Capt. Evans, R.N.

4. The most desirable position in which to build an iron ship for the N hemisphere, is, with her head to the south, and *vice versa.*

5. Every iron ship, after launching, should be kept with her head in an opposite direction to that in which she was built, *i.e.,* while she is being equipped for service.

6. In armour-plated vessels, the head should be kept in the same way, while the process is completing.

NAPIER'S DIAGRAM.

The introduction of iron vessels within the past few years, both in the Navy and Mercantile Marine, has rendered the question of deviation most important, and demanded at the hands of the skilful navigator an intelligent acquaintance with the principles which have been recently brought to light in connection with the care and management of a standard compass. We have said, (see page 8), that the deviations obtained by cross bearings or otherwise, for every position of the ship's head, are tabulated and kept for use when at sea. But, by an ingenious arrangement, known as Napier's Diagram, the deviation may be so laid down that the true course may be obtained from the steering course at sight, and without calculation.

Let a line 18 inches in length be divided into 32 equal parts for points of the compass, and subdivided into 360°. Such a line may be regarded as the edge of a compass card, flattened out to form a straight line. Around each point in succession let the chord of 60° be taken from Gunter's scale and a circle described. On the edge of this circle measure off from the vertical line arcs of 60° right and left, and join those points to the centre of the circle, taking care to let left-handed lines be dotted or imperfect. It will be evident that all these lines must cut the vertical line at an angle of 60°, forming equilateral triangles on either side. This gives the skeleton form of Napier's diagram, which is now sold prepared, so that its construction by the seaman is not needed. The table of deviations having been prepared the curve is thus laid down :—

In the preceding diagram, from N measure off 2° on the right hand from the scale on the vertical line, to be reckoned on the dotted or imperfect line. From NbE measure 5°, from NNE 7° &c., &c. When the deviations change to west, (as they do at SSE), measure on the left hand side of the vertical line, and by joining the points thus obtained, a curve is produced generally symmetrical in form, and of the shape in the accompanying figure.

From this curve may be seen, at sight, the magnetic from the true

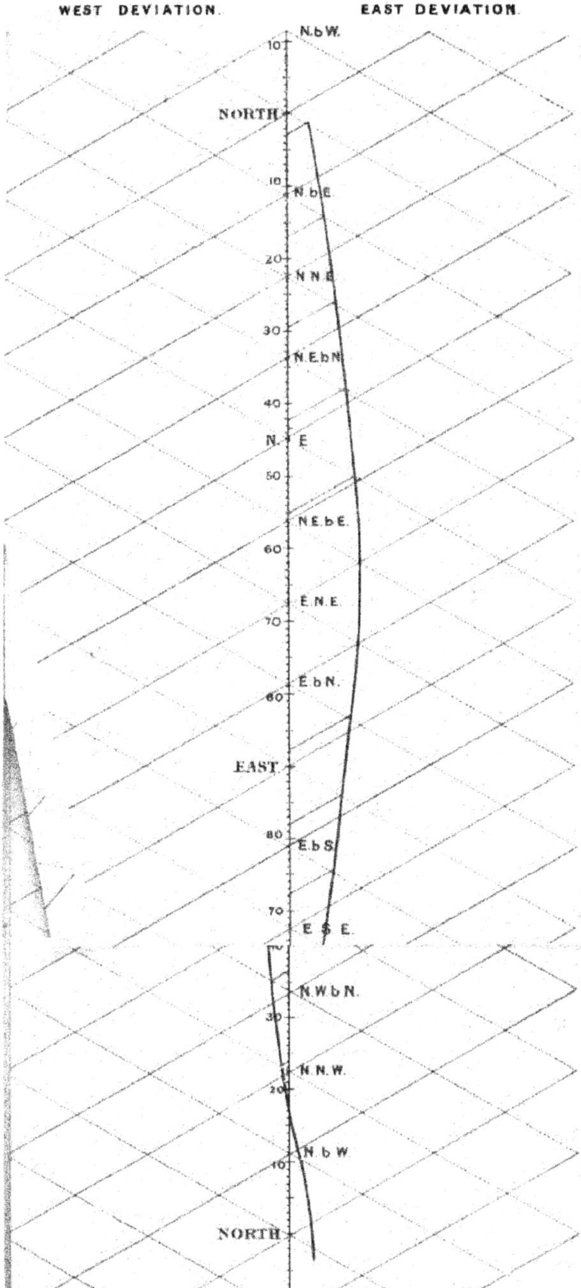

… quoted, may be

…ted."

…ue course. Start
…curve meets it is
…line which will
…about E ¼ N.
…mperfect one), we
…and *vice versa*.

…the line produced
…is actually forced
…or nearly so, there
…he sea rises there
…to the judgment
…he trim and form
…neral, ships when
…ter make little or
…n sailing close to

…vessel be at A,
…time, and let the
…influence of the
…that time, but by
…her own motive
…nor C, but, (by a
…mechanics), she
…he parallelogram,

…es by reference to
…ble of deviations."

course and *vice versa*, and the following rhyme, elsewhere quoted, may be useful to assist the memory,—

> "From Compass Course Magnetic Course to gain,
> Depart by dotted and return by plain,
> But, if you seek to steer a Course allotted,
> Take plain from Chart, and keep her head on dotted."

Thus, let the magnetic course be EbN, to find the true course. Start from EbN on the dotted line till the point where the curve meets it is reached. From that point, return parallel to the plain line which will bring to about 3° above E. The true course is, therefore about E ¼ N.

Or, better still, if we have a magnetic course, (*i.e.* an imperfect one), we start upon the imperfect or dotted line to find the true; and *vice versa.*

Leeway

May be correctly defined as the angle contained between the line produced of the ship's keel, and the line of direction in which she is actually forced through the water by the wind. When the wind is aft, or nearly so, there is no leeway. But, when the wind draws forward and the sea rises there is always more or less leeway. The amount must be left to the judgment of the experienced seamen alone ; it is dependent upon the trim and form of the vessel. It is commonly supposed that, in general, ships when sailing close to the wind and going fast through the water make little or no leeway, but it is now well known that all ships, when sailing close to the wind, make leeway more or less.

Fig. 2.

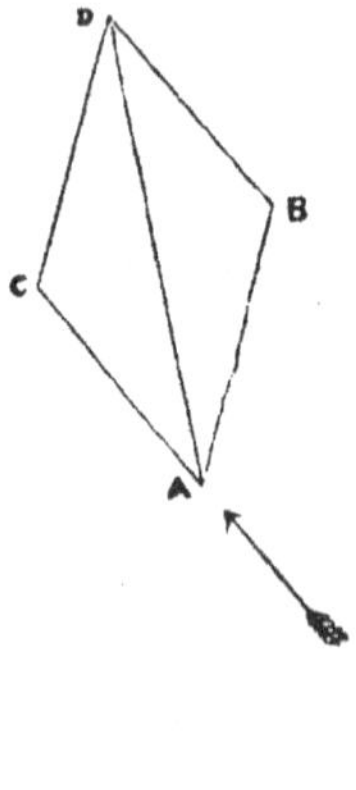

In the adjoining figure, let a vessel be at A, intending to sail to B in a given time, and let the wind be from S.E. If left to the influence of the wind alone, she would reach C in that time, but by the combined force of the wind and her own motive power, she will arrive neither at B nor C, but, (by a well-known principle of elementary mechanics), she will travel along the diagonal of the parallelogram, ABCD, and reach D.

Correction of Courses.

We now proceed to illustrate the Correction of Courses by reference to the preceding compass table, and also the subjoined " table of deviations."

(I.) When the compass course is given, to find the true course,—

When the vessel has been swung (see page 7), the deviation for each point of the compass is thus tabulated:—

TABLE OF DEVIATIONS FOR GIVEN POSITIONS OF THE SHIP'S HEAD.

Direction of Ship's Head.	Deviation of Compass.	Direction of Ship's Head.	Deviation of Compass.
N	2°·45′ E	S	3°·0′ W
N b E	4·57	S b W	4·20
NNE	7·30	SSW	5·0
NE b N	9·0	SW b S	6·7
NE	10·0	SW	7·0
NE b E	10·55	SW b W	7·27
ENE	10.40	WSW	7·50
E b N	9·55	W b S	8·20
E	8·50	W	8·50
E b S	7·15	W b N	8·10
ESE	5·35	WNW	6·50
SE b E	3·40	NW b W	5·40
SE	1·50	NW	4·50
SE b S	0·20 E	NW b N	3·20
SSE	0·56 W	NNW	1·40 W
S b E	2·20	N b W	1·10 E

EX. 1.—Given the Compass Course, WSW, variation $2\frac{1}{2}$ pts. W, leeway 2 pts., wind S.E.: to find the true course.

RULE WCE.

The above letters will enable the student to remember, that, all easterly variation and deviation, in this case, is called right, and westerly left, as indicated by the positions of E and W with reference to C (Compass Course).

Here, Compass Co, WSW.

Now, entering the table at WSW, the direction of the ship's head, we find the deviation corresponding to be 7° 50′ W = ¾ pt. nearly, (for 1 pt. = 11°.15′), then

	pts.		qrs.	
Comp Co	6	.	0	*r*S (right of South).
Devⁿ·			3	*l* (allowed to left) of C. in WCE.
	5	.	1	*r*S
Varⁿ·	2	.	2	*l* (allowed to left) of C in WCE.
	2	.	3	*r*S
Leeway	2	.	0	*r* (allowed to right) or SW ¾ W.
True Co	4	.	3	*r*S

In allowing for leeway, the vessel going WSW and the wind blowing from the SE, places her on the port tack, hence the leeway is right.

Ex. 2.—Given Compass Co ENE, variation 3¼ pts E, leeway 2 pts, wind N, find the true course.

Here

	pts.		qrs.	
Comp Co	6	.	0	*r*N
Devⁿ·	1	.	0	*r*
	7	.	0	*r*N
Varⁿ·	3	.	1	*r*
	10	.	1	*r*N
Leeway	2	.	0	*r*
	12	.	1	*r*N
Subtract from	16	.	0	(being over 8 pts)
True Course	3	.	3	*l*S (name reversed)
Or	SE ¼ S			(see note).

EXAMPLES FOR EXERCISE.

Comp. Co.	Wind.	Varⁿ·	Leeway.	True Course.
WNW ½ W	N b E	2½ pts. W	2	SW ¼ W
ESE ¼ E	N	1½ E	1½	SSE ¾ E
W	NE	2¼ W	¾	SW ¼ W
SW ¾ S	E	1½ E	¼	SW ¼ W
NNE ¼ E	W	1¼ W	1¼	NE b N
NW ¼ W	E	2½ E	¾	NW b N

Note.—When the course is over 8 pts. or 90°, it must be subtracted from 16 pts. or 180°, care being taken to change the name, thus, 10 pts. *r* N becomes 6 pts. *l*S., *i.e.* N 10 pts. E becomes S 6 pts. E, and so on.

Example Worked in Degrees.

Given Comp. Co. WNW½W, wind NbE, var. 2½ pts. W, leeway 2 pts find the true course.

Here, Comp. Co. = 6½ pts.	=	73° 7′ *l*N
Dev.	=	7 30 *l*
		80 37 *l*N
Var. = 2½ pts.	=	28 7 *l*
		108 44 *l*N
Leeway = 2 pts.	=	22 30 *l*
		131 14 *l*N
Subtract from		180 0
True Course	=	48 46 *r*S (see note)

or

S 48° 46′ W

The examples in the preceding table may be similarly worked, and the result, (more accurate), may be compared with the answer in points.

True Course Given to Find Compass Course.

Rule ETW

From which we gather that easterly variation and deviation are allowed to the left, and westerly to the right.

Given the True Co. SSW½W, variation 2 points W, to find the compass course.

In this case the variation must be first applied to the true course to obtain an approximate magnetic course, and with this latter the table must be entered for the deviation when the approximate compass course is thus found. The table should be, strictly, entered a second time for the deviation, but, for all practical purposes, one entry will be found sufficient. Thus,

		pts.		qrs.	
True Co. SSW½W	=	2	.	2 *r*S	
Variation	=	2	.	0 *r*	(allowed to right of T in ETW.)
Approx^te. Mag^c. Co.	=	4	.	2 *r*S	
Dev^n. with this Co. from table	=			3 *r*	(allowed to right of T in ETW.)
Compass Course		5	.	1 *r*S	

or

SWbW¼W.

The above worked in degrees.

True Co. SSW½W = 28°.7′ rS
Variation 2 pts. = 22 .30 r
Approx^te. Mag^c. Co. = 50 .37 rS (or 4½ pts.
Dev^n. with this Co. from table = 7 .13 r
Comp course 57 .50 rS
or
S 57°.50′ W

EXAMPLES FOR EXERCISE.

True Course.	Variation.	Comp. Course.
ENE½E	2 pts. W	E¼N
WSW¾W	2½ „ E	SWbW
EbN½N	1¾ „ W	E½N

DEFINITIONS IN NAVIGATION.

We have noticed that the earth is nearly a sphere and that it has two motions, one on its own axis from west to east which it accomplishes in 24 hours, the other round the sun in 365¼ days.

The extremities of the axis are called the **poles**: that to which we in Europe are nearest is called the **north pole**, the other the **south.**

Meridians are great circles passing round the earth through both poles.

The Equator is a great circle passing round the earth at an equal distance from both poles.

Parallels of latitude are less circles parallel to the equator: those which are situated 23½° north and south of the equator forms the boundaries of the **Tropics.** That parallel of latitude 23½° north is called the **Tropic of Cancer,** the southern one the **Tropic of Capricorn.** A line or more properly a curve drawn from one place to another, and cutting all the meridians it meets at the same angle, is called the **Rhumb line** or **Loxodromic curve.**

The latitude of any place is that part of the meridian intercepted between the place and the equator.

The difference of latitude between two places, is that portion of any meridian intercepted between the parallels of latitude drawn through the places.

The longitude of a place is that portion of the equator which lies between a given meridian, (called the **first meridian**), and the meridian over the place. Longitude is thus measured east and west of the meridian of Greenwich, and never exceeds 180°.

The **difference of longitude** between two places is that part of the equator intercepted between the meridians of those places.

The **meridian distance** between two places, is the portion of a parallel of latitude intercepted between them.

The **departure** which a ship makes in sailing on a rhumb line, is the sum of all the intermediate meridian distances supposing the rhumb line to be divided into an indefinitely small number of parts.

The **course** of a ship is the angle which the rhumb line on which she sails makes with any meridian it cuts, and the number of miles sailed on that rhumb line is called the **nautical distance**.

The **middle latitude** is the algebraical mean between two given latitudes: *i.e.*, if both north or south it is half their sum, but if of contrary names half their difference.

A **geographical or nautical mile** is 6079 English feet, for the earth being a sphere of 7916 English miles in diameter, or 24869 in circumference, it follows that a nautical mile is the 21,600th part of 360° or 6079 feet.

Meridional difference of latitude is the value in minutes of arc of the line on a Mercator's chart into which the true difference of latitude has been expanded.

Meridional parts is the increase in the lengths of small portions of the meridian as laid down on a Mercator's chart.

Leeway is the angle between a ship's actual course, and her intended course, caused by the action of the wind.

PREPARATORY PROBLEMS.

I.—To find the diff lat and diff long between two places.

Rule.—When the given latitudes and longitudes are of the same name, subtract the less from the greater, opposite names add them. When the sum of the longitudes exceeds 180°, subtract from 360°.

Ex. 1.—Find the D lat and D long between A and B,

Lat. A	49° 30′ N	Long. A	112° 30′ W
Lat. B	61 10 N	Long. B	118 10 W
	11 40		5 40
	60		60
D lat.	700′ N	D long.	340′ W

named N and W because in going from A to B, the ship would sail in those directions.

Ex. 2.—Find the d lat and d long between A and B.

Lat A	61° 10′ N	Long A	106° 30′ W
Lat B	52 30 N	Long B	99 10 W
	8 40		7 20
	60		60
D lat	520′ S	D long	440′ E

Ex. 3.—Find the d lat and d long between A and B.

Lat A	1° 45′ N	Long A	110° 44′ E
Lat B	2 10 S	Long B	92 40 W
	3 55		203 24 W
	60		360 0
D lat	235′ S		156 36
			60
		D long	9396′ E

Examples for Exercise.

Find the d lat and d long in the following Examples:—

Lat. A.	Lat. B.	Long. A.	Long. B.	D lat.	D long.
51° 30′ N	62° 10′ N	84° 30′ E	93° 16′ E	640′ N	526′ E
72 45 S	81 10 S	112 40 E	115 17 E	505 S	157 E
59 20 N	31 2 S	74 20 E	114 10 W	5422 S	10290 E
37 20 S	59 12 S	71 30 W	112 40 W	1312 S	2470 W
55 12 N	62 40 N	122 10 E	112 30 W	448 N	7520 E
6 30 N	5 19 S	107 21 E	110 42 E	709 E	201 E

II.—The latitudes and longitudes from being given, and the d lat and d long, to find the latitude and longitude in.

Ex. 1.—If a ship sail from lat 49° 58′ N, long 5° 20′ W towards the SW till her d lat is 210 miles, and d long 150 miles, find her lat and long arrived at.

D lat =	6,0)21,0	D long =	6,0)15,0
	3° 30′ S		2° 30′ W
Lat from	49 58 N	Long from	5 20 W
Lat in	46 28 N	Long in	7 50 W

N.B.—The D lat and D long are named according to the course sailed.

Ex. 2.—If a ship sail from lat 2° 30′ S, long 15° 30′ E towards the SW, till her d lat is 310 miles, and her d long 1,006 miles, find her lat and long arrived at.

D lat = 6,0)3,10		D long = 6,0)100,6		
	5° 10′ S		16° 46′ W	
Lat from	2 30 S	Long from	15 30 E	
Lat in	7 40 S	Long in	1 16 W	

EXAMPLES FOR EXERCISE.

Lat. from.	Long. from.	Course.	D lat.	D long.	Lat. in.	Long. in.
59° 18′ N	39° 10′ W	SE	210	240	55° 48′ N	35° 10′ W
62 12 S	40 24 E	NW	130	430	60 2 S	33 14 E
51 27 S	110 30 W	NE	270	160	46 57 S	107 50 W
42 10 N	90 30 E	SE	410	500	35 20 N	98 50 E
16 30 S	111 40 W	NE	160	420	13 50 S	104 40 W
72 20 N	39 15 E	SW	140	350	70 0 N	33 25 E

III.—Error of Log Line and Sand Glass.

The rate at which a vessel sails is estimated by means of the Log Line. The log is a piece of wood in the form of the sector of a circle. The lower part is loaded with lead so as to make it sit uprightly in the water, and with its flat surface towards the ship it offers sufficient resistance to the water as to be practically considered stationary. A line of about 150 fathoms (or 900 feet) is attached to this, and is divided into certain equal spaces called **knots**. At 10 or 12 fathoms from the log-ship a rag of bunting is placed, which marks off what is termed the "stray line" which lets the log-ship go clear of the vessel before time is counted. A sand glass is also used to mark the time in which a portion of the line is run out. The length of a knot on the line really depends on the number of seconds the log-glass measures; it bears the same ratio to the nautical mile, that the time of the glass does to one hour, or we may say,

$$\frac{\text{The feet in 1 knot}}{\text{The feet in 1 mile}} = \frac{\text{The seconds the glass runs}}{\text{The seconds in one hour.}}$$

A **knot** is the 120th part of a mile, and a **fathom** is generally considered as the **tenth** part of a knot. In practice, however, the knot is taken as fifty feet, and the time is measured by a glass supposed to run out in thirty seconds. When the glass is run out the knots between the ship and the extremity of the stray line indicate the distance run in that interval, and thus the hourly rate of sailing is known.

Now the action of the sea may serve to contract the length of the log line, or that of the atmosphere may affect the time in which the sand glass runs out. The following formulæ are therefore suggested to correct these sources of error:—

Let K represent the true length of a knot (50 ft.)
„ k „ actual „ „
„ d „ erroneous distance measured
„ D „ true distance.

$$\text{Then } K : k :: d : D, \text{ or } D = \frac{kd}{K}$$

whence the error of the line is corrected for.

But that of the glass yet remains,—

Let T represent the time the glass should take to run out, (viz., 30^s)
„ t „ „ it actually takes
„ D „ the distance corrected for line
„ x „ „ „ line and glass.

Then

$$t : T :: D : x$$

$$\text{or } x = \frac{T.D}{t}.$$

Example.—If the length of a knot be 52 feet and the distance run by log line 87 miles, the glass running out in 34 seconds, what is the real distance run?

$$(1)\ D = \frac{kd}{K} = \frac{52 \times 87}{50} = 90.48$$

$$(2)\ x = \frac{T.D}{t} = \frac{30 \times 90.48}{34} = 79.84$$

Ans. 79.84 miles.

Examples:

	Distance by Log.	Length of Knot.	Seconds by Glass.	True Distance.
1.	95	49	27	103.3
2.	136	51	32	130
3.	317	48	31	294.2
4.	420	47	29	408.4
5.	500	52	26	600
6.	370	53	33	356.6

Plane Sailing.

In Plane Sailing, the portion of the earth traversed is considered to be a plane surface, the meridians being represented as parallel to each other, and the parallels of latitude as straight lines crossing them at right angles.

Navigation may be divided into two great problems, viz., 1.—Given the latitudes and longitudes of two places, to find the course and distance from one to the other, and 2.—Given the latitude and longitude of a place, and the course and distance, to find the latitude and longitude in. Now if we attempt to solve these problems on the principles of plane sailing, we must use a plane chart, in which the mer. dist. is equal everywhere to the D. long; the results thus obtained would be very erroneous, unless the vessel sailed near the equator. If however, we introduce besides these, another element called departure, then up to a certain point the problems of navigation may be solved on the principles of plane sailing.

Fig. 3.

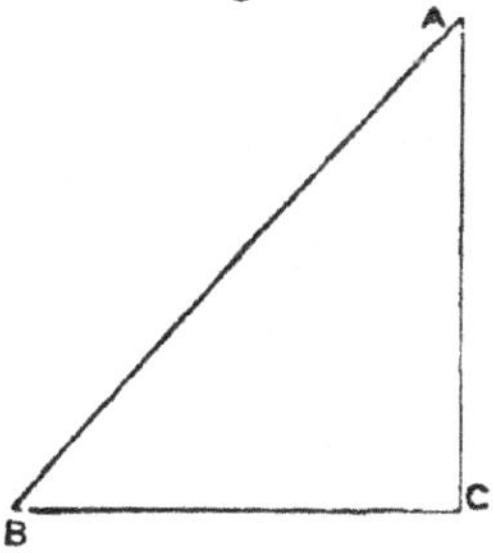

Let A and B represent respectively the place started from and that arrived at, AC the meridian of A; then AB will be the **Distance** sailed, AC the **Diff Lat**, and the perpendicular BC on AC will be the **Departure** made, the angle A the **Course.**

These elements are connected together by the Trigonometrical canons, viz.:—

$$\frac{BC}{AB} = \sin A \therefore BC = AB \sin A$$

$$\text{or dep} = \text{dist sin co (1)}$$

$$\frac{AC}{AB} = \cos A \therefore AC = AB \cos A$$

$$\text{or d lat} = \text{dist cos co (2)}$$

$$\frac{BC}{AC} = \tan A \therefore BC = AC \tan A$$

$$\text{or dep} = \text{d lat tan co (3)}$$

$$\frac{BA}{AC} = \sec A \therefore BA = AC \sec A$$

$$\text{or dist} = \text{d lat sec co (4)}$$

$$\frac{AC}{BC} = \cot A \therefore AC = BC \cot A$$

$$\text{or d lat} = \text{dep cot co (5)}$$

$$\frac{AB}{BC} = \operatorname{cosec} A \therefore AB = BC \operatorname{cosec} A$$

$$\text{or dist} = \text{dep cosec co. (6)}$$

Assuming the student to be acquainted with elementary trigonometry, whence the preceding formulæ are derived, it must be noted that these should be thoroughly committed to memory, so that they may be at his ready command.

The following may be readily seen from them, and may in a measure assist the work, viz :—

(1) Course and dist being given,—we have sin and cos co, for dep and d lat.

(2) Course and d lat being given tan and sec co for dep and dist.

(3) Course and dep being given cot and cosec co for d lat and dist.

Ex. I.—A ship sails from lat 51° 25′ N, SSW$\frac{1}{2}$W, 210 miles: find her departure and lat. in.

(1) *By Construction.*

From Gunter's scale, on the line of chords, with the compasses, measure the chord of 60°, and with this as radius describe a circle as in the figure:

Fig. 4.

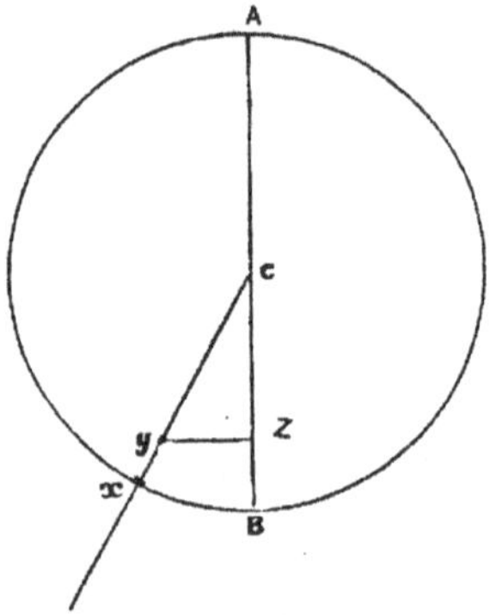

Draw a vertical line ACB, to represent the meridian sailed from. Then from the line of rhumbs measure $2\frac{1}{2}$ points, and lay it off on the circumference from B towards the west, as the course is SSW$\frac{1}{2}$W, as B*x*. Join C*x* and produce it if necessary: then BC*x* is $2\frac{1}{2}$ points (the course). From a scale of equal parts measure along C*x* the dist = 210 C*y*. Draw *yz* perpendicular to CB. Then from the scale from which C*y* was measured, *yz* the dep will be found to be 99 miles, and C*z* the d lat 185 miles.

Similarly the course and another side being given the figure may be laid off, and the unknown parts measured. When two sides are given, *e.g* *yz* and *zc* the course *ycz* is determined by measurement.

(2) *By Inspection.*

Enter the traverse table, with course $2\frac{1}{2}$ points, and dist 210: then opposite this will be found dep = 99, and d lat = 185.2. This is the readiest method of working questions in plane sailing. The table may be entered for any two given parts, when the remaining ones may be seen by inspection.

(3) *By Calculation.*

Dep = dist sin co
Log dep = log dist + log sin co − 10
dist = 210 = 2.322219
Co = 2½ pts. sin = 9.673387
Dep 99 = 1.995606

D lat = dist cos co
Log d lat = log dist + log cos co − 10
dist = 210 = 2.322219
Co = 2½ pts. cos = 9.945430
D lat 185.1 = 2.267649
3°5′.1 S
Lat from 51.25 N
Lat in 48.19.9 N

Ex. 2.—A ship sails in the SE quarter, from lat 49° 58′ N till her dep is 210 miles and d lat 314, find her course and dist and lat in.

(1) *By Construction.*

Fig. 5.

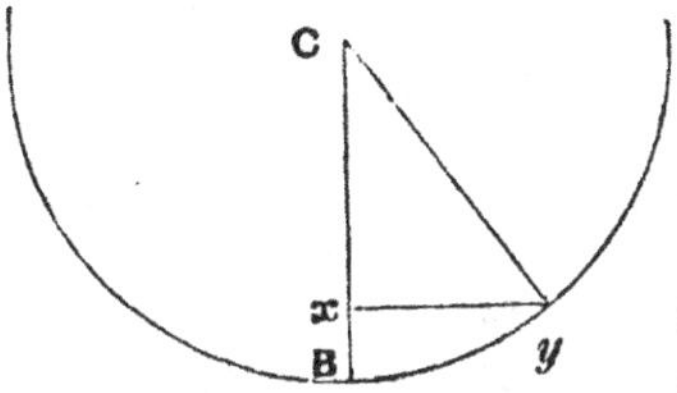

With the chord of 60° describe a circle. Draw the line CB to represent the meridian. From a scale of equal parts lay off on CB, the D lat=314; let this be C*x*. Measure the dep *xy* 210 perpendicular to C*x*, then *y* will fall on the circle, and joining C*y*, we have B*y* the measurement of the course, 3 pts, and C*y* the dist.

(2) *By Inspection.*

Enter the traverse table with dep 210, and d lat 314, when the course will be found 3 points, and distance 378.

(3) *By Calculation.*

$\frac{\text{Dep}}{\text{D lat}}$ = tan co
Log dep+10 log d lat=log tan co
Log dep = 12.322219
Log d lat = 2.496930
33°46 tan 9.825289
Course S 33° 46′ E

Dist = d lat sec co
Log dist = log d lat + log sec co − 10
Log d lat = 2.496930
Log sec co = 10.080238
377.7 = 2.577168
Dist 377.7

EXAMPLES FOR EXERCISE.

1. If a ship sail from Cape Clear, (lat 51° 25′ N), 240 miles SSW¾W: find her dep and lat in.

Ans. Dep 123.4 miles lat: 47° 59′ 6″ N.

2. A ship sails from lat 47° 20′ S till she arrives in lat 49°10′ S, having made 240 miles of westerly departure: find her course and distance.

Ans. Co S 65° 22′ W. Dist 263·9.

3. If a ship leave lat 42° 10′ S, and sail WSW¾W, until she arrive in lat 47°.20′ S: find her dist and dep.

Ans. Dep 1238. Dist 1276.

4. Sailed from lat 49° 58′ N. WSW, till our dep was 100 miles, find the dist run and lat arrived at.

Ans. Dist 108.2 Lat 49° 16′ 35″ N.

5. If a ship sail between the south and west, from lat 51°.25′ N, till she reaches lat 49° 10′ N, 200 miles, find her course and dep.

Ans. Co S 47° 32′ W. Dep 147.5.

6. A ship from lat 42° 54 N, sails SW½W, till her dep is 115 miles; find her dist and lat in.

Ans. Dist 148.8 Lat 41° 19′ 38″ N.

7. On Jan 8th at 8 A.M., Halifax bore from us NW¼N, 14 miles, we then sailed for 1½ hours on an opposite course, at 7 miles an hour; find the dep and dist run.

Ans. Dep 16.45. Dist 24.5.

8. If a ship sail from lat 49° 10′ S, SW¾W, 5 miles an hour, in what time will she reach the parallel of 51° S.

Ans. 36.9. hours.

9. If a ship sail from Funchal, lat 32° 37′ N between the south and west till her diff lat is 110 miles and dep 110 miles: find her course, distance, and lat arrived at.

Ans. Co S 45° W; dist 155.5; lat 30° 47′ N.

10. A ship sails between the north and east until her dep and d lat are equal: find her course.

Ans. Co N 45° E.

11. If a vessel sail until her distance made is double her d lat; what is her course?

Ans. 60°.

12. If a ship's distance sailed be double the dep made, what is the course?

Ans. 30°.

13. A ship from lat 42° 30′ N reaches lat 47° 20′ N, having made 76 miles of departure, find the dist without the aid of tables.

Ans. Dist 299.9 miles.

14. If a ship sails 200 miles S, having made 100 miles westerly departure, from lat 20° N, find the course and lat in.

Ans. Co S 30° W. Lat 17° 6′ 48″ N.

Traverse Sailing.

Various causes, *e.g.* contrary winds, rocks, shoals, &c., prevent a vessel continuing for any length of time upon one course; she is therefore compelled to sail upon a series of zigzag courses, and the crooked line thus described is termed a Traverse. Traverse Sailing is the method of resolving into one single mean course and distance this series of irregular courses; this being done, the problem is treated as one in plane sailing, or as if the vessel had sailed in one line without changing her course, from her starting point to her destination.

Ex.—A ship sails from lat 51° 25′ N, SSE¼E 16, ESE 23, SWbW½W 36, W¾N 12, and SEbE¼E 41 miles: find her mean course and dist (or course and dist made good), and latitude in.

Figure.

The courses and distances may be laid off as shewn in plane sailing, and the problem solved by construction as in the examples given. It may also be solved by inspection.

In the accompanying figure the vessel may be represented as going from A to D,E,F,G,B. Then the triangle ACB represents the plane sailing triangle, or as if she had sailed directly from A to B.

Fig. 6.

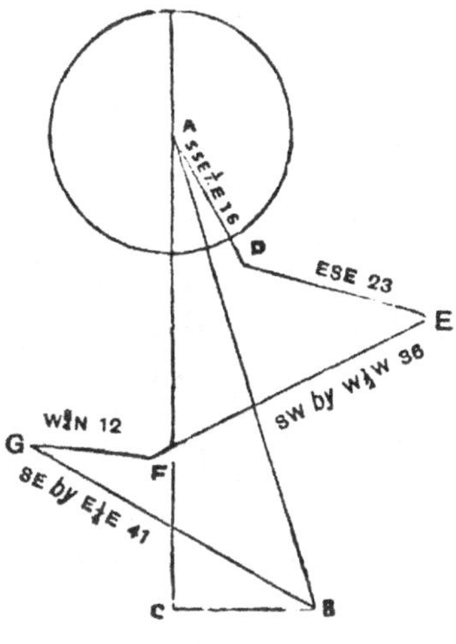

TRAVERSE TABLE.

Courses.	Dist.	Diff. Lat.		Departure.	
		N	S	E	W
SSE$\frac{1}{4}$E	16	...	14.5	6.8	...
ESE	23	...	8.8	21.3	...
SWbW$\frac{1}{2}$W	36	...	17.0	...	31.8
W$\frac{3}{4}$N	12	1.8	...	...	11.9
SEbE$\frac{1}{4}$E	41	..	21.1	35.2	...
		1.8	61.4	63.3	43.7
			1.8	43.7	
			60)59.6	19.6 E dep	

D lat 0.59.36 S
Lat from 51.25. 0 N
Lat in 50.25.24 N

$$\text{Tan co} = \frac{\text{dep}}{\text{d lat}}$$

Log dep + 10 = 11.292256
D lat = 1.775246
Co : S 18° 12′ E tan 9.517010

Dist = d lat sec co
D lat = 1.775246
Sec co = 10.022289
Dist = 62.7 1.797535

Ex. 2.—If compass courses be given with the variation, the true courses must first be obtained before placing them in the Traverse Table.

If a ship from lat. 38° 42′ N sail on the following compass courses, viz., SSW $\frac{1}{4}$ W 21, WSW $\frac{3}{4}$ W 18, and S $\frac{1}{2}$ W 20 miles, Var. 20° W, find her true course, dist and lat in.

CORRECTION OF COURSES.

(1)	(2)	(3)
Comp co 28° 7′ rS	Comp co 75° 56′ rS	Comp co 5° 37 rS
Var 20 0 l	Var 20 0 l	Var 20 0 l
True Co S 8 7 W	S 55 56 W	S 14 23 E

Courses.	Dist.	Diff Lat.		Departure.	
		N.	S.	E.	W.
S 80° 7′ W	21	...	20.8	...	2.9
S 55 56 W	18	...	10.1	...	14.9
S 14 23 E	20	...	19.4	4.8	...
			6,0)5,0.3	4.8	17.8
					4.8
					Dep 13.0 W

D lat 0.50.18 S
Lat from 38.42. 0 N
Lat in 37.51.42 N

$\text{Tan Co} = \frac{\text{Dep.}}{\text{D lat.}}$	Dist = D lat Sec Co.
Dep = 11.113943	D lat = 1.701568
D lat = 1.701568	Sec Co = 10.014026
S 14° 29′ W tan = 9.412375	51.95 = 1.715594
Ans. Co S 14° 29′ W.	Dist. 51.95 miles

EXAMPLES FOR EXERCISE.

1. Sailed from lat. 47° 20′ S on the following courses:—SSW ½ W 18, WSW ½ S 19, W 25, S 14, S b E ½ E 24 miles; find the lat, and course and distance made good.

Ans. Lat 48° 21′ 54″ S, course S 34° 58′ W, dist 5.53.

2. If a ship sail on the following compass courses, from lat 49° 58′ N, viz., S b W 22, WSW ½ W 19, WNW 27, S 35, S b E ½ E 19, var 22° 30′ W; find her lat, and course and distance.

Ans. Lat 48° 37′ 24″ N, course S 8° 28′ W, dist 81.5.

3. Sailed from lat 16° 30′ S, on the following:—N b E ¾ E 10 NE, ½ E 14, NNW ½ W 27, W b S 30, SSW ½ W 25; find the lat, course and distance.

Ans. Lat 16° 15′ 54″ S, course N 70° 27′ W, dist 42.13.

4. If a ship sail from lat 42° 30′ S, on the following:—NNW ¾ W 21, W ½ S 17, W 39, W ½ N 15, WNW ½ W 10; find her lat, course and distance.

Ans. Lat 42° 9′ 18″ S, course N 77° 13′ W, dist 93.5.

5. On leaving Cape Clear (lat 51° 25′ N), the land bore from us NE ¾ N 20 miles; we then sailed as follows: SSW ½ W 14, S 27, S b E ¼ E 19, SW ¾ S 17, W ½ N 28 miles, find our lat; and the course and distance made.

Ans. Lat 50° 0′ 12″ N, course S 31° 28′ W, dist 99.42.

6. On January 19th, Lisbon (lat 38° 42′ N), bore from us NE ½ E 15 miles; since then we sailed WSW ½ S 21, WNW ½ W 17, S ½ E 15, S b W ¼ W 20, and NNE 15 miles; find our latitude on January 20th at the same time, also the course and distance made.

Ans. Lat 38° 6′ 12″ N, course S 50° 55′ W, dist 56.79.

7. Sailed from lat 42° 30′ N, on the following compass courses, var 1¾ pts W, viz.: SE ½ S 24, S b W ½ W 19, S ¾ W 27, SE b S 17, S 14 miles; find the lat, course and distance.

Ans. Lat 41° 8′ 30″ N, course S 29° 8′ E, dist 92.84.

8. A ship from lat 51° 10′ S sails on the following courses: NE ½ N 17, NE ½ E 23, E b N ½ N 15, E ½ S 27, E 15 miles, var 23° 27′ E; find the lat, course and distance.

Ans. Lat 51° 3′ 12″ S, course S 85° 39′ E, dist 89.65.

PARALLEL SAILING.

When a vessel sails either due east or west, her path lies on a parallel of latitude; thus her latitude remains the same, while her longitude is continually changing. This method is called "Parallel Sailing," which may be considered as the link between plane and spherical sailing. The distance run on a parallel is called the "Meridian Distance," (see definitions), and it bears a constant ratio to the difference of longitude. In latitudes not higher than 5°, and where the meridian distance does not exceed 300 miles, the departure may be used for the difference of longitude, the resulting error scarcely exceeding a mile. When the methods of finding the longitude were not so reliable as now, it was the custom to find the latitude of the place of destination, and then run down its parallel due east or west until it was reached.

Example 1.—A ship sails from long 110° 25′ W, 200 miles west, on the parallel of 36° 35′ S; find her long in.

Here D long = mer dist sec lat
,, = 200 × sec 36° 35′
Log 200 = 2.301030
Log Sec = 10.095289
D long 249.1 = 2.396319

6,0)24,9.1

D long	4°	9′	6″	W
Long from	110	25	0	W
Long in	114	34	6	W

Example 2.—What is the length of a degree on the parallel of 50°?

Here one degree at the equator = 60 miles
= d long
and mer dist = d long cos lat
= 60 cos 50°
Log 60 = 1.778151
Cos 50° = 9.808067
38.57 = 1.586218

Therefore a degree in lat 50° is 38.57 nautical miles.

EXAMPLES FOR EXERCISE.

1. If a ship sails from long 75° 33′ W, 100 miles west on the parallel of 57°, find her longitude in.

Ans: Long 78° 36′ 36″ W.

2. A ship sails from long 15° W, to long 17° 30′ W on the parallel of 30° find the distance.

Ans: Dist 130 miles.

3. If a ship sail 200 miles from long 30° W, reaching long 35° W, on what parallel does she sail?

Ans: Lat 48° 11′.

4. How far must a ship sail on the parallel of 50°, to change her longitude 2° 30′.

Ans: Dist 96.42 miles.

5. At what rate does the earth revolve in the latitude of Greenwich 51½°.

Ans: 560.2 miles an hour.

6. In what lat does a ship change her long 1°, when she has only sailed 35 miles

Ans: Lat 54° 23′.

7. A is carried 2½ times as fast as B by the earth's rotation: B is 20° North of A: find their latitudes.

Ans: 57° 38′ and 77° 38′.

8. Two places whose meridian distances are 100 and 150 miles, have 20° diff lat: find their latitudes.

Ans: 38° 37′, and 58° 37.

9. The meridian distance of two places is $(a - b)$; and of the two others in a less latitude $(a + b)$; the sum of the latitudes is 90°: find an expression for the d lat.

$$\text{Ans: Tan } (l - l') = \frac{b}{a}$$

10. The meridian distance of the two places is double that of two other places in a higher latitude: and the latitudes are as 1 : 2; show that

$$\text{Cos } l = \frac{1 + \sqrt{33}}{8}$$

MIDDLE LATITUDE SAILING.

We have hitherto considered cases in which either the latitude or the longitude is changed, the former coming under the head of Plane Sailing, the latter Parallel Sailing: we now have to consider the change both of latitude and longitude, under the title Middle Latitude Sailing; which is founded upon supposition, that the arc of the parallel of latitude midway between two places, is equal to the departure between them, which is not strictly true, but sufficiently so to answer the purposes this sailing seeks to fulfil. The method is useless in high latitudes when the course is small, and also when the places lie in opposite hemispheres ; but it may be safely used in low latitudes when the places are on the same side of the Equator, and when the course is more than 45° or 4 points, in which case the effect of the assumed error is least.

Example 1.—Sailed from lat 51° 25′ N, long 9° 29′ W, WSW ½ W, (var 2½ pts W), 210 miles, find our latitude and long in.

Compass course 6½ pts *r*S
Varⁿ. 2½ *l*
True course 4 *r*S

To find Latitude.

D lat = Dist cos co
Dist 210 log 2.322219
Co: 4 pts cos 9.849485
6,0)14,8.4 = 2.171704
D lat 2° 28′ 24″ S
Lat from 51 25 0 N
Lat in 48 56 36

Lat in 48° 56′ 36″
Lat from 51 25 0
2)100 21 36
Mid lat 50 10 48

To find Longitude.

D long = Dist sin co sec mid lat
Dist 210 = 2.322219
Sin co 4 pts = 9.849485
Sec mid lat = 10.193594
6,0)23,2 = 2.365298
D long 3° 52′ W.
Long from 9 29 W.
Long in 13 21 W.

N.B.—This should be used only when the course exceeds 4 points.

Example 2.—Required the compass course and distance from A to B.

Lat A 51° 25′ N
Lat B 52 37 N
D lat 72 N = 1 12
2)104 2
Mid lat 52 1

Long A 9° 29′ W
Long B 12 30 W
3 1
60
D long 181 W

Var 2½ pts W

To find Course.

$$\text{Tan co} = \frac{\text{D long cos mid lat}}{\text{D lat}}$$

D lat 181 = 2.257679
Mid lat cos = 9.789180
12.046859
D lat 72 = 1.857332
57° 7′ tan 10.189527

True co N 57° 7′ W. . *l*
Varⁿ 28 7 W. . *r*
Approx mag co 29 0 . . *l*
Dev (Table page 12.) 2 30 W. *r*
Comp. co. N 26 30 W

The devⁿ. is applied as in example page 14.

To find Distance.

Dist	=	d lat sec true course
D lat 72	=	1·857332
True co 57° 7′ sec	=	10·265256
132·6	=	2·122588

Ans. Comp co N 26° 20′ W; dist 132·6 miles.

Example 3.—Required the compass course and distance from A to B.

Lat A 31° 30′ S	Long A 129° 10′ W	Var $2\frac{1}{2}$ pts W.
Lat B 31 30 S	Long B 117 12 W	

Since the ship does not change her latitude, she sails on a parallel, and her true course is 8 points east, and the distance run is her meridian distance, hence—

Mer dist	=	d long cos lat
	=	718 cos 31° 30′
D long 718	=	2·856124
31° 30′ cos	=	9·930766
Dist 612 2	=	2·786890

To find Comp Course.

True co S	90°	0′	E	*l*
Var	28	7	W	*r*
Approx mag co	61	53		*l*
Dev	7	38	W	*r* (Table page 12)
Comp co S	54	15		E

EXAMPLES FOR EXERCISE.

1. Sailed from lat 42° 40′ S, long 25° 20′ W, SW $\frac{1}{2}$ W 212 miles: find lat and long in.

Ans. Lat 44° 54′ S, long 29° 7′ W.

2. Sailed from lat 49° 58′ N, long 6° 16′ W, due west, var $2\frac{1}{2}$ pts W, dev $\frac{1}{2}$ pt E, 150 miles, find the latitude and longitude.

Ans. Lat 49° 0′ 36″ N, long 9° 49′ 18″ W.

3. Sailed from lat 53° 34′ N, long 0° 7′ E, ENE $\frac{1}{2}$ E, var $2\frac{1}{2}$ W, dev $\frac{1}{2}$ pt W, 100 miles: find lat and long in.

Ans. Lat 54° 51′ 18″ N, long 1° 55′ 42″ E.

4. Required the compass course and dist from A to B, var $2\frac{1}{2}$ pts W.

Lat A 50° 25′ N	Long A 12° 30′ W
Lat B 47 10 N	Long B 7 50 W

Ans. Comp co S 13° 40′ E, dist 268.4.

5. Find the compass course and dist from A to B, var $1\frac{3}{4}$ pts E.

Lat A 14° 30′ S	Long A 15° 30′ W
Lat B 10 20 S	Long B 27 12 W

Ans. Comp co N 80° 49′ W, dist 730.

6. Find the compass course and dist from A to B, var $2\frac{1}{2}$ pts E.

Lat A 51° 30′ N	Long A 12° 40′ W
Lat B 51 30 N	Long B 27 10 W

Ans. Comp co S 69° 31′ W, dist 542.

7. Required the compass course and dist from A to B, var $2\frac{1}{4}$ pts W.

Lat A 27° 20′ S	Long A 37° 40′ W
Lat B 25 12 S	Long B 42 12 W

Ans. Comp co N 30° 41′ W, dist 275.

8. If a ship sail from lat 55° 40′ N, long 29° 39′ W, N 66° 13′ W, var $1\frac{1}{2}$ pts E, 200 miles, find her latitude and longitude in.

Ans. Lat 57° 51′ 12″ N, long 34° 15′ 48″ W.

9. Sailed from lat 33° 18′ S, long 72° W, S 59° 4′ W, var 16° E, dev 9° 25′ E, 54.61 miles; required the lat and long.

Ans. Lat 33° 23′ S, long 73° 5′ W.

10. Required the compass course and distance from A to B, var $2\frac{1}{2}$ pts W.

Lat A 15° 20′ S	Long A 4° 10′ E
Lat B 19 40 S	Long B 1 20 W

Ans. Comp Co S 86° 53′ W, dist 306.5 miles.

Middle Latitude by Construction.

Fig. 7.

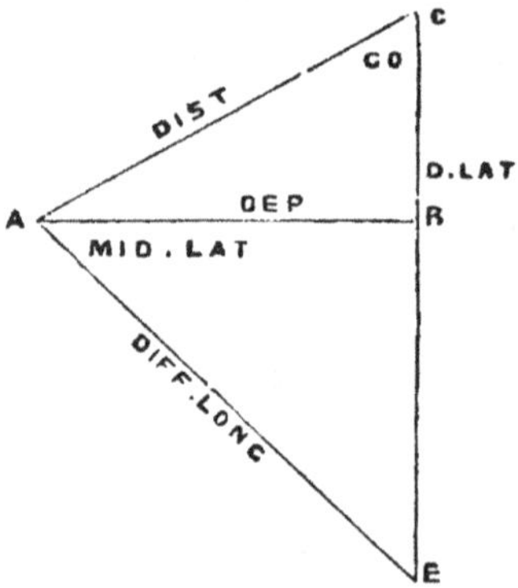

By the formula D long = Dep sec m lat we recognise the relation that exists between the hypothenuse and one of the other sides of a right angled triangle with reference to the adjacent angle.

Hence in the triangle ABE, AE = D long: AB = Dep, and the angle EAB = mid lat, while the triangle ABC contains the elements of plane sailing.

Now, if mid lat = 50°, and dep = 179, by taking the chord of 60° round A and laying off mid lat = 50° measured from AB towards E, and AB 179, we shall find AE the diff long = 280 nearly.

By Inspection.

Enter the traverse table with the mid lat as a course, and the D long as distance, then under the heading D lat we have the departure; or more correctly enter the table with the co mid lat (90°—mid lat), and the D long, then dep comes in its own column.

Mercator's Sailing.

We are indebted for this method partly to Gerard Kauffman, the Latin equivalent of whose name is Mercator, and who was born in East Flanders in 1512. Though the originator of the method, he was unacquainted with its value, and until 1594 the law which regulates the increase of the degrees of latitude with their distance from the equator, (the peculiarity of Mercator's projection), was unobserved, when Mr. Edward Wright made it known.

Let us first notice, however, the principles of Mercator's projection, under the heading of

Mercator's Chart.

Imagine the earth to be touched at every point of the equator by a circumscribed cylinder of white paper, and conceive the eye to be placed at the centre of the globe. If the meridians of the globe be delineated on the paper cylinder, as they appear to the eye, they will become straight lines parallel to each other, and the parallels of latitude will be represented by circles equal to the equator.

Let the cylinder be now unrolled into a plane surface: the equator has now become a straight line: the meridians have become straight lines at right angles to it, and parallel to each other; and the parallels of latitude also straight lines everywhere equal to the equator.

Now a chart to be of service to the navigator must represent the precise relation that exists between the minute division of the meridian, and the minute divisions of a parallel near to some point on the globe. Thus, if on the globe the length of a minute of the parallel of 60° be one half the length of a minute of a meridian, then on a chart that same relation should hold good. Or, if in lat 70°, the length of a minute of a meridian be about thrice the length of a minute of the parallel, the same relation must obtain on the chart.

But these relations, the above projection does not accurately give, and recourse is had to a more exact method, still keeping to the idea of the developed cylinder.

On the globe, if a minute of the equator be taken as the limit of measurement, then a minute of any parallel of latitude must be multiplied by the secant of the latitude to be made equal to a minute of the equator. But, as was shown above, the parallels have become equal to the equator, (and therefore to the meridians), consequently they have become multiplied by the secants of their latitudes: and in order that the proper relations may still be preserved at all points, the minutes of the meridian must be increased in the same degree, that is, in the ratio of the secants of the latitudes of those points.

The length, therefore of a minute of a meridian at any particular point on a Mercator's Chart is found by multiplying a true minute of the meridian

by the secant of the latitude of that point; and the sum of all the increased minutes contained in a portion of the meridian between two parallels is termed the "meridional difference of latitude." A table containing all these secants for every minute of latitude is called a table of "meridional parts."

It will be obvious that a chart or map constructed on such a framework will give an entirely false idea of the relative areas of countries in different latitudes. At the equator alone, portions of the globe will appear as they are on the globe, while the more distant a country is from the equator, the more is its area exaggerated: the shape is all that remains true. This disadvantage however, is more than counterbalanced by the precious properties it possesses for the purposes of navigation.

The shortest distance between two points on the globe is along the arc of a great circle joining them; but, as an arc of a great circle cuts the meridians at different angles, to sail on it, it would be necessary to change the course continually.

The curve, however, which does cut meridians at the same angle, and which, for distances not immense, is little longer than the arc of a great circle, is called a "rhumb line;" and, as only straight lines can cut parallel lines at the same angle, it follows that a rhumb line must be delineated on a Mercator's Chart by a straight line.

The chief value of a Mercator's Chart is that **the angle which the rhumb line makes with a meridian, is exactly equal to the corresponding angle measured on the globe**; thus the actual course is faithfully depicted upon the chart.

To draw a Mercator's Chart.

Take a horizontal line to represent the lowest parallel on the chart, divide and subdivide it as may be necessary for degrees of longitude, and take this for a line of measurement.

Draw a straight line perpendicular to this from its extremity, to represent the extreme meridian.

Find the mer diff lat between the lowest parallel and every other to be laid down on the chart, divide these differences by 60 for degrees. Then taking the quotients from the bottom line, or line of measures, apply them from the bottom parallel on the extreme meridian: thus all the parallels to be contained on the chart will be determined. The land may then be sketched in, together with the position of rocks, shoals, &c., and a compass, for the measurement of courses from point to point.

To find the course from point to point.

Lay the edge of a parallel rule over the places, and move the rule parallel to this position until it passes the centre of a compass, where the course will be seen.

To find the distance from point to point.

Find the middle latitude between the points.

(*a*) If the places are in the same parallel or in the same latitude, take half their distance, and apply it on the graduated meridian on both sides of the parallel, and the distance will be found.

(*b*) If the places are not in same latitude, lay a scale over both, and taking half their distance apart, measure this from the middle parallel between them, on either side of it, then the distance will be determined on the graduated meridian.

To find the latitude and longitude of a place.

Find the perpendicular distance of the place from any convenient parallel and measure on the graduated meridian from the point where the parallel cuts it, on the same side as the places, which will give its latitude.

From the latitude and longitude of a place, to find its position on a chart.

Lay a scale over the parallel of the given latitude, and measure on a graduated parallel the distance of any convenient meridian from the given longitude: apply this along the edge of the scale from the point where the meridian measured cuts the scale, in the same direction that the longitude lies from the meridian, then the point to which the distance reaches will be the position of the place.

MERCATOR'S SAILING.

In "plane sailing" a right angled triangle is constructed similar to the numerous elementary triangles formed by a rhumb line cutting the meridians on the globe, and having its sides equal in length to the sum of the corresponding sides of the small triangles. Now if these small triangles of the globe be projected on a Mercator's Chart, and another similar triangle be constructed as in "plane sailing," with its sides equal in length to the sum of the corresponding sides of the projected triangles, we shall obtain a triangle similar to the first or "plane sailing" triangle.

Fig. 8.

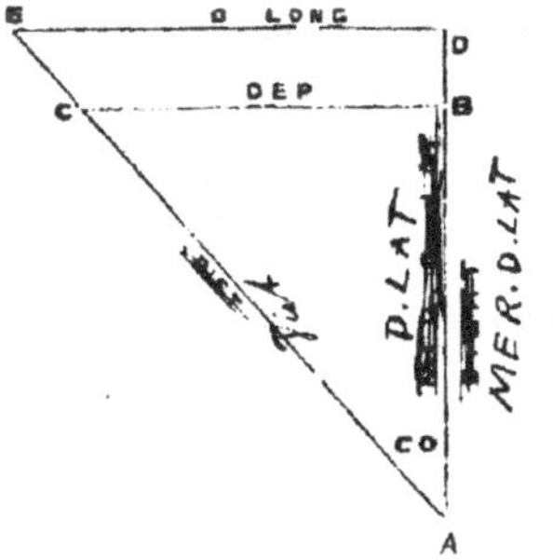

Let ABC and ADE be the two similar triangles, having the common angle at A. Then AB is the diff lat, AD the meridional diff lat, BC the departure, DE the diff long, AC the dist sailed, AE the projected distance, and the angle A the course, then,—

$$\frac{ED}{DA} = \tan A$$

$$i.e.\ \frac{\text{D long}}{\text{Mer d lat}} = \text{tan course.}$$

Also,

ED : DA :: CB : BA
D long : mer d lat :: dep : d lat

Therefore,

$$\text{D long} = \frac{\text{mer d lat} \times \text{dep}}{\text{d lat}}$$

Example 1.—A ship sails from lat 49° 58′ N, long 12° 30′ W, SSW ½ W, 210 miles; find her lat and long in.

To find Latitude.

D lat = Dist cos co.

Log dist 210	= 2·322219
Co 2½ pts cos	= 9·945430
6,0)185·2	= 2·267649

D lat	3° 5′ 12″ S
Lat from	49 58 0 N
Lat in	46 52 48 N

Meridional parts.

For lat 49° 58′	= 3455
„ lat 46 52	= 3175
Mer d lat	= 280

To find Longitude.

D long = mer d lat tan co.

Mer d lat 280	= 2·447158
Tan co 2½ pts	= 9·727957
6,0)149·6	= 2·175115

D long	2° 29′ 36″ W
Long from	12 30 0 W
Lat in	14 59 36 W

N.B.—If the lat from and lat in be of contrary names, the mer d lat is found by adding the mer parts.

Example 2.—Required the compass course and distance from A to B; var 2½ pts W.

Lat A	47° 10 N
Lat B	40 25 N
	6 45
	60
D lat	405 S

Long A	52° 40′ W
Long B	50 20 W
	2 20
	60
D long	140 E

Meridional parts.

For lat 47° 10′	= 3202
„ „ 40 25	= 2642
Mer d lat	= 560

$$\text{Tan co} = \frac{\text{d long}}{\text{mer d lat}}$$

Log d long 140	= 12·146128
Log mer d lat 560	= 2·748188
True course S 14° 2′ E tan	9·397940

To find Compass Course.

True co	S 14° 2′ E . . *l*	
Var	28 7 W . . *r*	Dist = d lat, sec course.
Magnetic co	S 14 5 W . . *r*	Log d lat = 2·607455
Dev (page 12)	4 30 W . . *r*	Log sec co = 10·013159
Comp co	S 18 35 W	Dist 417.4 = 2·620614

By Construction.

Draw a vertical line, and round the point A take the chord of 60°. Measure on this vertical line AB the mer d lat = 560, and from B draw BC perpendicular to BA = D long 140.

Then the angle A, on the subtending arc will measure 14° (the course nearly).

Fig. 9.

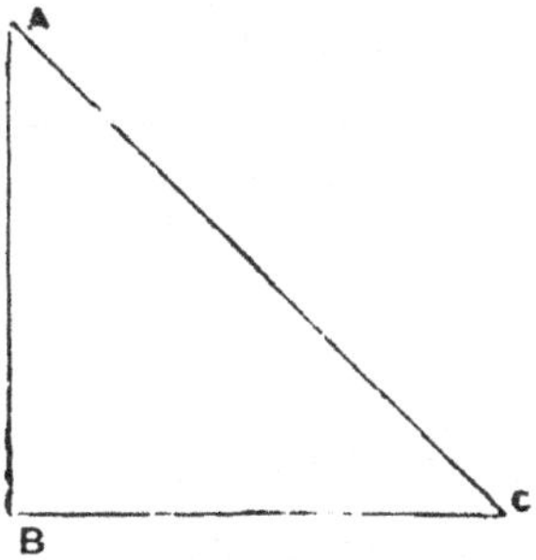

By Inspection.

Enter D long in the departure column of the Traverse Table, and mer d lat as the d lat, when the course will be found 14°.

EXAMPLES FOR EXERCISE.

1. A ship sails from lat 41° 20′ N, long 15° 20′ W, S ½ W 140 miles; find the lat and long in.

Ans. Lat 39° 0′ 42″ N, long 15° 38′ 0″ W.

2. Sailed from lat 42° 10′ N, long 17° 30′ W, N b W ¼ W, 200 miles; find the lat and long in.

Ans. Lat 45° 24′ N, long 18° 37′ 7″ W.

3. Sailed from lat 15° 20′ S, long 20° 45′ W, NNW ¾ W, 145 miles; find the lat and long in.

Ans. Lat 13° 15′ 42″ S, long 22° 2′ 19″ W.

4. Find the compass course and distance from A to B,

Lat A 49° 20′ N	Long A 15° 30′ W
Lat B 56 30 N	Long B 25 10 W
Var 1¾ pts W	

Ans. Comp co N 17° 54′ W; dist 554.

5. Find the compass course and distance from A to B.

Lat A 53° 18′ N	Long A 21° 30′ W
Lat B 60 27 N	Long B 16 30 W
Var $1\frac{1}{2}$ pts W	

Ans. Comp co N 28° 30′ E ; dist 459·1.

6. Required the compass course and distance from lat 49° 12′ N, long 12° 4′ W, to lat 36° 58′ N. long 25° 12′ W, var 22° 30′ E, dev 2° 15′ W.

Ans. Comp co S 17° 47′ W ; dist 932.

7. Sailed from lat 57° 20′ N, long 30° W, SW $\frac{3}{4}$ W, 147 miles ; find the lat and long in.

Ans. Lat 55° 41′ 17″ N ; long 33° 18′ 36″ W.

8. A ship sails from lat 34° 30′ S, long 25° 20′ W, SSW $\frac{1}{2}$ W, until she reaches lat 39° 10′ S ; find her long in.

Ans. Long 28° 27′ W.

9. A ship sails from lat 67° 30′ N, till her dep is 110 miles, and her d long 260 ; find her lat in.

Ans. Lat 62° 26′ N.

DAY'S WORK.

This is the most important problem in Navigation, and one, to which those we have considered tend. The Day's Work or Sea Journal is a register of various transactions connected with the navigation of the ship that have occurred during the 24 hours, *e.g.*, the periods when the course was changed ; the direction of the wind at the time ; the leeway, variation and deviation of the compass, together with the set and drift of the currents met with. The latitude and longitude of the ship by astronomical observations at any particular time are also inserted in the log, and the methods by which the position of the ship at the time was determined.

When a vessel leaves the land, the bearing of some known point or headland is observed by the Compass, and the eye of the practical seaman will determine readily and approximately its distance at that time. This is called "Taking the departure."

Each day at noon these results are transcribed from the log board on which the day's run is kept, into the log-book ; the courses are corrected for deviation, variation and leeway, and the results being entered in a Traverse Table, the latitude and longitude of the ship are obtained, which, compared with those deduced from astronomical observations, give the correct place of the ship. This method is known also as the "Dead Reckoning."

Various incidents that take place in the 24 hours, *e.g.*, sighting land, passing vessels, phenomena of the atmosphere, (lightning, squalls, &c.,) are usually inserted in the "remark" column.

In bad weather, it is usual, under a contrary wind and high sea, to avoid as much as possible being driven back; no more sail is carried than is necessary to prevent violent rolling; the tiller put over to leeward, brings the ship's head round to the wind, and thus her way through the water is lost. When her head falls off again from the wind, she makes fresh way again; and thus she comes up and falls off alternately. The middle point between that on which she comes up and falls off is taken as her apparent course, and the deviation, leeway, and variation being allowed from that point, the result is entered as a course in the Traverse Table, and the estimated drift of the ship through the water is taken as a distance. In correcting the current course, strictly speaking, the deviation ought to be allowed for, as well as the variation, but the direction of the ship's head will not generally be that of the current.

Example.

On a certain day, at noon, a point of land, lat 33° 2′ N, long 71° 41′ W, bore by compass E, ship's head NW), distant 10 miles; afterwards sailed as follows, &c.,

H.	K.	$\frac{1}{10}$ths	Courses.	Winds.	Lee-way.	Remarks.
1	4	3	WSW ½ W	E	½	
2	4	2				
3	5	7				
4	4	1				
5	5	5				
6	4	8				
7	6	1	W ½ N	ESE	½	Var$^{n.}$ 1¾ pts E
8	5	6				
9	4	9				
10	5	1				
11	4	3				
12	5	2				
1	4	1				
2	6	4				
3	5	6	SSW ¼ W	N	¾	
4	5	1				A current set the ship the last 5 hrs. 2½ miles an hour WSW, by compass.
5	6	6				
6	4	7				
7	4	3				
8	5	7				
9	6	6				
10	4	1	ESE	N	½	
11	4	7				
12	4	8				

The Deviation Table, before given, (page 12) must be entered with the given course, or direction of the ship's head, and remembering that $11^\circ\frac{1}{4} = 1$ pt., the nearest quarter point is to be sought from the given deviation: thus—

Departure Course.

	pts.	qrs.
Course (named reversed)	8 .	0 *r*S
Deviation for NW		2 *l*
	7 .	2 *r*S
Var^n.	1 .	3 *r*
	9 .	1 *r*S
Take from	16 .	0
True Co	6 .	3 *l*N

or WNW¾W

Course 2.

	pts.	qrs.
Comp Co	6 .	2 *r*S
Dev^n.		3 *l*
	5 .	3 *r*S
Var^n.	1 .	3 *r*
	7 .	2 *r*S
Leeway		2 *r*
True Co	8 .	0 *r*S

or W.

Course 3.

	pts.	qrs.
Comp Co	7 .	2 *l*N
Dev^n.		3 *l*
	8 .	1 *l*N
Var^n.	1 .	3 *r*
	6 .	2 *l*
Leeway		2 *r*
True Co	6 .	0 *l*N

or WNW

Course 4.

	pts.	qrs.
Comp Co	2 .	1 *r*S
Dev^n.		2 *l*
	1 .	3 *r*
Var^n.	1 .	3 *r*
	3 .	2 *r*S
Leeway		3 *l*
	2 .	3 *r*S

or SSW ¾ W

Course 5.

	pts.	qrs.
Comp Co	6 .	0 *l*S
Dev^n.		2 *r*
	5 .	2 *l*S
Var^n.	1 .	3 *r*
	3 .	3 *l*
Leeway		2 *r*
	3 .	1 *l*S

or SE ¾ S

Current Course.

	pts.	qrs.
Comp Co	6 .	0 *r*S
Var^n.	1 .	3 *r*
	7 .	3 *r*S

or W ¼ S.

The distances from the table are reckoned from the course to that preceding the next course.

Traverse Table.

Courses.	Dist.	D lat.		Dep.	
		N	S	E	W
pts.					
N $6\frac{3}{4}$ W	10	2·4	...	...	9·7
S 8 W	28·6	...	...	...	28·6
N 6 W	11·7	16·1	...	...	38·8
S $2\frac{3}{4}$ W	38·6	...	33·5	...	20·0
S $3\frac{1}{4}$ E	13·6	...	10·9	8·1	...
S $7\frac{3}{4}$ W	12·5	...	·6	...	12·5
		18·5	45·0	8·1	109·6
			18·5		8·1
		D lat	6,0)26·5	dep	101·5 W

0° 26′ 30″ S
Lat from 33 2 0 N
Lat in 32 35 30
2)65 37 30
Mid lat 32 48 45

D long = dep sec mid lat
Log dep = 2·006466
Log sec mid lat = 10·075509
6,0)12,0·7 = 2·081975
D long 2° 0′ 42′ W
Long from 71 41 0 W
Long in 73 41 42 W

EXAMPLES FOR EXERCISE.

1. A point of land in latitude 59° 30′ S, long 30° 40′ E, bore, on a certain day, at noon, SSE, distant 18 miles, ship's head being NNE; afterwards sailed as follows: find her latitude and longitude next day at noon.

H.	K.	$\frac{1}{10}$ths	Courses.	Winds.	Lee-way.	Remarks.
1	4	5	NE b E	NNW	1	P.M.
2	6	0				
3	7	0				
4	5	5				
5	6	5				
6	3	5				Pts.
7	5	0	SW	SE	$\frac{3}{4}$	Var 1¼ E.
8	4	5				
9	6	0				
10	7	5				
11	6	0				
12	6	5				
1	6	0	S b W	W	1	A.M.
2	7	0				
3	5	5				
4	4	5				
5	6	0				A current set the
6	6	5	NNE	NNW	1¾	ship the last two
7	7	0				hours NE by com-
8	6	5				pass 4½ miles per
9	8	0				hour.
10	7	5				
11	6	0				
12	7	5				

True Courses.—N 18, S 7¾ E 33, S 5¼ W 36, S ¾ W 29, N 5 ¾ E 49, N 5¼ E, 9.

Lat 59° 35′ 12″ S. Long 32° 18′ 12″ E.

2. On a certain day, at noon, a point of land in latitude 44° 30′ S, long 56° 30′ E bore by compass NE b E, 15 miles, ship's head being SW; find the lat and long next noon.

H.	K.	$\frac{1}{10}$ths.	Courses.	Winds.	Lee-way.	Remarks.
1	4	5	N ½ E	E	1¼	P.M.
2	5	3				
3	3	5				
4	6	0				
5	4	5				
6	3	2				
7	6	5	WSW	S	1½	Pts. Var 1¾ E.
8	4	6				
9	3	3				
10	4	5				
11	7	0				
12	5	2				
1	5	5	SW ½ W	NW	1¾	A.M.
2	4	0				
3	5	7				
4	6	3				
5	4	5				A current set the
6	5	4	NE ½ E	SE	¾	ship the last five
7	6	7				hours NW by
8	4	4				compass 2½ miles
9	5	5				per hour.
10	3	6				
11	6	0				
12	3	4				

True Courses.—S 6 W 15, N 1¼ E 27, N 7½ W 31, S 3¾ W 26, N 6½ E 35, N 2¼ W, 13.

Lat 44° 3′ 48″ S, long 55° 51′ 11″ E.

3. On a certain day, at noon, a point of land in lat 63° 40′ S, long 54° 30′ W, bore by compass NE b E 15 miles, ship's head being ENE; afterwards sailed as follows; find our position next day at noon.

H.	K.	$\frac{1}{10}$ths.	Courses.	Winds.	Lee-way.	Remarks.
1	3	5	N $\frac{3}{4}$ W	ENE	1	P.M.
2	4	4				
3	6	0				
4	4	5				
5	7	0				
6	5	5				Pts.
7	6	0	SSW	E b S	$\frac{3}{4}$	Var $2\frac{1}{2}$ W.
8	2	5				
9	5	5				
10	4	6				
11	7	0				
12	5	5				
1	6	0	NE b E	SE	$2\frac{1}{4}$	A.M.
2	5	5				
3	3	5				
4	4	0				
5	2	6				A current set the
6	3	5				ship the last 8
7	4	5	SW b S	WNW	$1\frac{1}{2}$	hours SE by com-
8	5	5				pass, $2\frac{1}{2}$ miles an
9	3	6				hour.
10	4	0				
11	5	5				
12	6	5				

True Courses.—S $3\frac{1}{2}$ W 15, N $4\frac{1}{4}$ W 31, S $\frac{1}{4}$ E 31, N $1\frac{1}{4}$ E 25, S $1\frac{1}{2}$ E 30, S $6\frac{1}{2}$ E 7.

Lat 64° 8′ 12″ S, long 54° 51′ 36″ W.

4. At noon, on a certain day, a point of land in lat 55° 58′ S, long 9° 30′ W, bore by compass ENE ½ E 12 miles, ship's head being SSW: find our lat and long next noon.

H.	K.	$\frac{1}{10}$ths.	Courses.	Winds.	Lee-way.	Remarks.
1	6	4	NW ½ W	SE	0	P.M.
2	7	5				
3	3	4				
4	5	5	NNW ¾ W	S	1	
5	6	6				
6	7	5				Pts.
7	8	5				Var 1¾ E.
8	9	4				
9	5	4	N	SSW	¼	
10	10	4				
11	11	5				
12	9	5				
1	6	1				A.M.
2	5	8				
3	6	6				
4	4	4	NE	WSW	½	
5	6	2				A current set the
6	5	9				ship the last 7
7	5	8				hours, WSW by
8	4	7				compass 2½ miles
9	5	5	NNW	NE	¾	per hour.
10	4	4				
11	5	6				
12	4	6				

True Courses.—S 7¾ W 12, N 3¼ W 17, N 38, N 2 E 55, N 7¼ E 27, N 1 W 20, S 7¾ W 18.

Lat 53° 54′ S, Long 9° 24′ W.

5. A point of land in lat 48° 32′ N, long 29° 36′ E, bore by compass SE 20 miles, ship's head N, find the position at the following noon.

H.	K.	$\frac{1}{10}$ths.	Courses.	Winds.	Lee-way.	Remarks.
1	9	2	W	E	0	P.M.
2	3	4				
3	4	6				
4	8	3				
5	7	5				
6	6	7				
7	6	6	SW b W	S	$\frac{3}{4}$	Var $3\frac{1}{2}$ pts E
8	5	1				
9	9	1				
10	9	4				
11	12	3				
12	7	2				
1	7	3				A.M.
2	4	2	W b S $\frac{1}{2}$ S	N	$\frac{3}{4}$	
3	6	5				
4	7	4				A current set the
5	4	2				ship from 9 p.m.
6	9	3				to 6 a.m., E b S
7	9	3				by compass 3
8	10	3	N	E	1	miles an hour.
9	11	3				
10	12	2				
11	11	2				
12	9	4				

True Courses N $\frac{1}{2}$ W 20, N $5\frac{1}{4}$ W 40, N $7\frac{1}{4}$ W 57, N $7\frac{1}{2}$ W 41, N $2\frac{1}{2}$ E 54 S $3\frac{1}{2}$ E 27.

Lat 49° 48′ N, long 27° 17′ E.

6. A point of land lat 50° 29′ N, long 0° 28′ W, bore from us NNE 20 miles, ship's head W, find the position at the end of the next 24 hours.

H.	K.	$\frac{1}{10}$ths.	Courses.	Winds.	Lee-way.	Remarks.
1	8	2	S ¼ W	E	1	P.M.
2	6	2				
3	7	2				
4	5	3				
5	5	3				
6	5	3				
7	6	4				Var 1¾ pts E.
8	9	2	ESE	S b W	¾	
9	7	6				
10	11	2				
11	11	4				
12	11	5	N b W	NE	½	
1	10	3				A.M.
2	10	4	SE	NE ½ E	¾	
3	10	5				
4	9	2				
5	8	1				A current set the ship the last 6 hours, SSE by compass, 2 miles an hour.
6	9	3	NW b W	NE b N	1	
7	7	2				
8	6	2				
9	5	4	ESE	NE		
10	4	7				
11	4	6				
12	6	5				

True Courses S 3 W 20, S 2¾ W 44, S 4½ E 39, N ¼ E 22, S 1½ E 38, N 4¾ W 23, S 3¾ E 21, S ¼ E 12.

Lat 48° 42′ N. long 0° 21′ W.

7. July 20th, at noon, Cape Comorin (lat 8° 5′ N, long 77° 30′ E, bore by compass NNE, ship's head S, dist 15 miles; afterwards sailed as follows: find the lat and long next day at noon.

H.	K.	$\frac{1}{10}$ths.	Courses.	Winds.	Lee-way.	Remarks.
1	5	2	WSW ½ W	E b S	¼	P.M.
2	6	3				
3	4	1				
4	5	8				
5	6	2				Variation of Compass 2¾ pts E.
6	5	7				
7	4	9				
8	5	3				
9	6	4	SSW	N	¼	
10	4	0				
11	4	2				
12	5	3				
1	5	5				A.M.
2	4	1				
3	6	6				
4	4	2	W ½ S	NE	½	A current set the ship the last 4 hours ESE ½ E by Compass 1.5 miles an hour
5	5	1				
6	6	7				
7	6	3				
8	5	6				
9	5	2				
10	6	1	W	E b S	0	
11	4	6				
12	4	5				

Ans. Lat 7° 46′ N, long 75° 26′ E.

8. September 17th, at noon, St. Michael, Azores, (lat 37° 48′ N, long 25° 8′ W), bore by compass NE ½ E, ship's head W b S, dist 12 miles; afterwards sailed as follows: find the lat and long in on September 18th, at noon.

H.	K.	$\frac{1}{10}$ths.	Courses.	Winds.	Lee-way.	Remarks.
1	6	1	SSW ½ W	E	¾	P.M.
2	6	2				
3	6	3				
4	5	7				
5	5	9				
6	6	8				Variation of
7	4	6				Compass 1½
8	5	1	W ½ S	SE	¾	pts E.
9	6	4				
10	4	2				
11	6	7				
12	5	5				
1	6	4				A.M.
2	5	6				
3	4	7				
4	6	8				A current set
5	5	1	S ½ E	WSW	½	the ship the
6	5	6				last 4 hours
7	4	2				W ½ S by
8	6	7				Compass 2
9	5	5				miles an hour
10	5	4	S	NE	½	
11	4	6				
12	4	6				

Ans. Lat 36° 44′ 6″ N, long 27° 20′ W.

9. October 12th, at noon, Rio Janeiro, (lat 22° 54′ 42″ S, long 43° 9′ W), bore by compass W ½ S, ship's 'head S, dist 17 miles; afterwards sailed as follows: find the lat and long in next day at noon.

H.	K.	$\frac{1}{10}$ths.	Courses.	Winds.	Lee-way.	Remarks.
1	7	2	SE ½ E	W	½	P.M.
2	6	8				
3	7	9				
4	5	3				
5	6	0				
6	7	1				Variation of Compass 1½ pts E.
7	6	1				
8	7	6	E	SW	½	
9	6	2				
10	7	2				
11	6	7				
12	5	4				
1	5	6				A.M.
2	6	5				
3	7	3				
4	6	1	ENE	SW	¾	A current set the ship the last 5 hours W½S by Compass 2½ miles an hour.
5	6	8				
6	6	7				
7	5	6				
8	6	7				
9	5	2	SE	WSW	½	
10	6	1				
11	5	7				
12	6	3				

Ans. Lat 24° 5′ 30″ S; long 40° 49′ 24″ W.

10. December 12th, at noon, Callao, (lat 12° 4′ S, long 77° 13′ 42″ W), bore by compass ENE, ship's head W, distant 14 miles; afterwards sailed as follows: find the lat and long in next day at noon.

H.	K.	$\frac{1}{10}$ths.	Courses.	Winds.	Lee-way.	Remarks.
1	4	2	WSW ½ W	NE	½	P.M.
2	5	1				
3	4	7				
4	5	3				
5	6	2				Variation of Compass 1½ pts E.
6	5	8	W	SE	½	
7	4	1				
8	6	6				
9	5	5				
10	5	4				
11	6	2				
12	4	7				
1	6	1				A.M.
2	4	7	S ½ W	E	½	
3	5	6				
4	4	2				A current set the ship the last 4 hours W ½ S 2 miles an hour.
5	5	7				
6	4	8				
7	6	1	SSW ¼ W	E	½	
8	6	6				
9	5	9				
10	5	2				
11	6	9				
12	4	8				

Ans. Lat 12° 49′ 42″ S; long 79° 19′ 12″ W.

GREAT CIRCLE SAILING.

On a plane surface, the shortest distance between two points is the straight line which joins them : on a globe or sphere the least distance is the arc or portion of the great circle contained between them. At first sight, it might be imagined that to sail on a rhumb line would be to run on the shortest distance from place to place, and to demonstrate this mathematically would be more difficult than to do so practically. Let a terrestial globe be taken and two points be selected on it : let the rhumb line distance be measured by a string : afterwards bring the two places down to the wooden horizon (which is a great circle), by tilting the globe, it will then be found that the arc of the great circle contained between them is less than the distance measured by the string. A ship, therefore, sailing on a great circle, steers for her port as if it were in sight. But as a great circle (unlike the rhumb line of the sailings), cuts every meridian it meets at a different angle, it would be necessary, in order that a vessel should sail upon it, to change her course continually : this being practically impossible, recourse is had to approximate great circle sailing, in which the vessel may be said to sail upon the sides of a many sided plane figure, (a polygon), which almost coincide, when taken together, with the great circle. Several points on the great circle track joining the place left and that to be arrived at, are selected : and if these be not placed very widely apart, the sum of the rhumb lines thus sailed upon, will not fall far short of the actual great circle distance.

The advantage of great circle sailing is most conspicuous in high latitudes, and its practical utility is seen when treated as an auxiliary to rhumb sailing.

This method may be comprised under five headings :

1. The distance on the great circle.
2. Latitude of the vertex.
3. Longitude of the vertex.
4. To find a succession of points on the great circle track.
5. To compute the course and distance from point to point.

Any great circle will cut the equator in two opposite points : the points midway between these will be those which have the greatest N or S latitude : such a point is called the vertex.

Example 1.—Required the distance on a great circle from Cape Clear (lat 51° 25′ N, long 9° 29′ W), to New York (lat 40° 42′ N, long 74° 1′ W), also latitude and longitude of the Vertex.

Lat A	40° 42′ N	Long A	74° 1′ W
„ B	51 25 N	„ B	9 29 W
Co lat A	49 18		2)64 32
Co lat B	38 35	½ d long	32 16

(1) *To find the Distance.*

½ d long	32°	16′	cos	9·927151
				2
				19·854302
Co lat A	49	18	sin	9·879746
Co lat B	38	35	sin	9·794942
	2)87	53		2)19·528990
$\frac{\text{Co lat A} + \text{co lat B}}{2}$ θ	43	56	sin	9·764495
	35	33		
Sum	79	29	sin	9·992643
Diff	8	23	sin	9·163743
				2)19·156386
½ Dist	22°	14′	sin	9·578193
		2		
	44	28		
	60			
Dist	2668 miles.			

(2) *Lat Vertex.*

Lat A	40°	42′	cos	9·879746
Lat B	51	25	cos	9·794942
D long	64	32	sin	9·955609
Dist	44	28	cosec	10·154595
Lat V	52	27	cos	9·784892

(3) *Long Vertex.*

Lat A	40°	42′	tan	9·934567
Lat V	52	27	cot	9·885765
D long	48	36	cos	9·820332

D long between A and V 48° 36′ E (∵ B is East of A).

Long A	74	1	W
Long V	25	25	W

Note.—Two facts must be borne in mind here.

1. A is the place of less latitude.
2. The D long is E or W according as B is E or W of A.

Example 2.—In the preceding example, what distance is saved by sailing on a great circle instead of a rhumb line.

Here computed dist = 2668 miles on great circle. Now, find the Compass Course and distance from A to B.

Lat A 40° 42′ N
B 51° 25′ N

Long A 74° 1′ W
B 9° 29′ W

This question will come under the head of Mercator's Sailing, the D lat being over 5°.

Mer parts.
For lat 40° 42′ = 2664
„ „ 51° 25′ = 3592
Mer d lat = 928

D long = 64° 32′
60
3872 miles
True d lat = 643 miles.

$\text{Tan co} = \dfrac{\text{D long}}{\text{M d lat}}$	Dist = D lat sec co
D long 3872 = 13·587935	D lat 643 = 2·808211
Mer d lat 928 = 2·967548	Sec co = 10·632341
Co N 76° 31′ E tan 10·620387	2757 = 3·440552

Then dist on rhumb line = 2757 miles
" " great circle = 2668
Dist saved = 89 miles.

It has been assumed in the above, that the reverse track has been taken, viz., New York to Cape Clear.

Example 3.—To find a succession of points on the preceding great circle track.

Rule.—To the log cos d long between vertex and point add the log tan lat vertex, and the result is log tan lat of point required: or

$$\text{Tan lat P} = \cos \text{d long} \times \tan \text{lat V}.$$

The longitudes 17°, 35°, vertex, and 46°, lie on this track; find the ship's lat while sailing on the great circle, over each of these longitudes.

Long V 25° 25′ W	Long V 25° 25′ W	Long V 25° 25′ W
Long A 17 0 W	Long B 35 0 W	Long C 46 0 W
D long 8 25	D long 9 35	D long 20 35

Lat A.	Lat B.	Lat C.
Lat V 52° 27′ tan 10·114235	Lat V tan 10·114235	Lat V tan 10·114235
Cos 8 25 9·995297	Cos 9° 35′ 9·993896	Cos 20° 35′ 9·971351
52° 9′ tan 10·109532	52° 4′ tan 10·108131	50° 36′ tan 10·085586

Therefore the points on the Great Circle are,

	Cape Clear.	A.	Vertex.	B.	C.	New York.
Lat	51° 25′ N	52° 9′ N	52° 27′ N	52° 4′ N	50° 36′ N	40° 42′ N
Long	9 29 W	17 0 W	25 25 W	35 0 W	46 0 W	74 1 W

Example 4.—To find the Course and Distance from point to point.

Course and Dist Cape Clear to A.

Lat 51° 25′ N	Long 9° 29′ W
Lat 52 9 N	Long 17 0 W
D lat 44 N = 0 44	7 31
2)103 34	60
M lat 51 47	D long 451 W

$$\text{Tan Co} = \frac{\text{D long cos m lat}}{\text{D lat}} \qquad \text{Dist} = \text{d lat sec co}$$

D long 451 =	2·654177		D lat 44 =	1·643453
M lat cos =	9·791436		Co sec =	10·807266
	12·445613		Dist 282·3 =	2·450719
D lat 44 =	1·643453			
Course N 81° 2′ W tan	10·802160			

Computing the other courses and distances similarly, we have as follows :—

	Course.	Dist.
From Cape Clear to A . . .	N 81° 2′ W	282·3 miles.
„ A to Vertex . .	N 86 40 W	309·5 „
„ Vertex to B . . .	S 86 15 W	351·6 „
„ B to C . . .	S 77 57 W	421·5 „
„ C to New York .	S 63 10 W	1316 „

EXAMPLES FOR EXERCISE.

1. Required the distance on a great circle from Rio Janeiro, (lat 22° 55′ S, long 43° 9′ W), to Cape Verd, (lat 14° 43′ N, long 17° 34′ W), also the lat and long of the Vertex.

Ans. Dist 2712 miles ; lat v 57° 9′ N ; long v 97° 48′ W.

2. Find the distance on a great circle from Cape Leeuwin (lat 34° 21′ S, long 115° 6′ E), to the Cape of Good Hope, lat 34° 22′ S, long 18° 29′ E), also the lat and long of the Vertex.

Ans. Dist 4566 miles, lat v 45° 47′ S, long v 66° 47′ E.

3. What distance would be saved by sailing on a great circle from Cape Coast Castle (lat 5° 5′ N, long 1° 14′ W), to Cape St. Roque (lat 5° 28′ S, long 35° 16′ W.)

Ans. Dist saved 10 miles.

4. Find the lats of the points A, B. C, D, on the great circle between Lisbon and Bermuda having given,

Lat Vertex 39° 41′ N, Long A 22° 30′ W, Long C 45° 20′ W.
Long „ 24 17 W, Long B 35 17 W, Long D 59 10 W.

Ans. Lat A 39° 40′; lat B 39° 9′ ; lat C 37° 45′; lat D 34° 14′.

5. Find the lats of the points A, B, on the great circle between Sierra Leone and Trinidad, having given,

Lat Vertex 10° 48′ N, Long A 27° 15′ W.
Long „ 51 50 W, Long B 42 26 W.

Ans. 9° 50′, and 10° 39′.

6. On the great circle between Brest (lat 48° 24′ N, long 4° 29′ W), and Cape Race (lat 46° 40′ N, long 53° 7′ W) select longitudes 20°, 37° and 45° West; find the latitudes of these points,

Ans. Lat A 50° 8′; lat B 49° 37′; lat C 48° 28′.

CURRENT SAILING.

As intimated under the heading Leeway, when a vessel sails subject to the action of a current, she does not reach the point to which her bow is directed, but falls off either to port or starboard, traversing the diagonal of a parallelogram. This line along which she travels is known in mechanics as the **Resultant** of the two **Component** forces which, acting together upon a body tend to produce motion in the direction of neither, but intermediate in direction. A current is named (unlike the wind), according to the direction to which it moves; thus, a current travelling SE would be a SE current; this direction is called the set of it. By the drift of a current is meant the distance over which the ship is carried by its action upon her; it is found by multiplying the known rate of the current by the number of hours during which the vessel is under its influence. This sailing is analogous to Traverse Sailing, the two courses instead of following in succession, being considered as taking place simultaneously, *e.g.*, as follows,—

1. A vessel sails 10 hours, on a SSW course 6 miles an hour, in a current setting E b N 2 miles an hour, find her true course and distance.

Course.	Dist.	D Lat. N.	D Lat. S.	Departure. E.	Departure. W.
SSW	60	...	55·4	...	23·0
E b N	20	3·9	...	19·6	...
		3·9	55·4	19·6	23·0
			3·9		19·6
			D lat 51·5 S		Dep 3·4 W

$$\text{Tan co} = \frac{\text{dep}}{\text{d lat}}$$

Dep 10·531479
D lat 1·711807
Co S 3° 46′ W tan 8·819672

Dist = d lat sec co

D lat 1·711807
Sec 10·000939
Dist 51·61 = 1·712746

2. A point of land was observed to bear due South; the ship then sailed from the point of observation ENE, 15 miles, when it bore SSW ½ W; find the distance from the point at each observation.

Fig. 10.

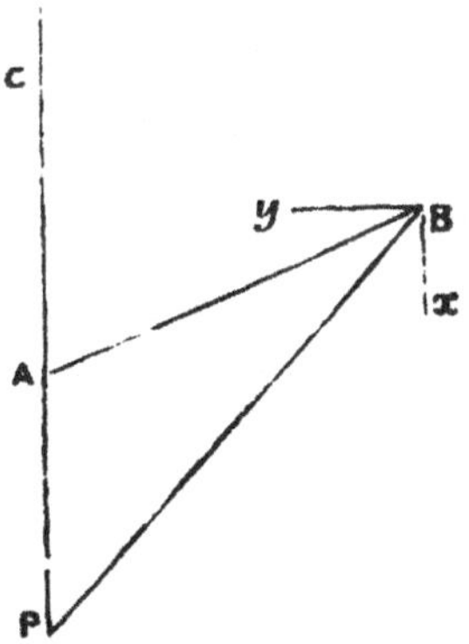

Here the angle CAB = ENE = 6 points
Therefore BAP = 16 pts − 6 pts
" " = 10 pts
And PBx = SSW [illegible] 2½ pts
And AB = 15

SOLUTION.

BAP = 10 pts
BPA = 2½ "
12½
Subtract fr. 16
ABP = 3½ pts

Then, As sin P : AB :: sin ABP : AP
or sin 2½ : 15 :: sin 3½ : AP
and AP = 15 . sin 3½ . cosec 2½
log 15 = 1·176091
3½ sin = 9·802359
2½ cosec = 10·326613
AP 20·18 = 1·305063

And, As sin P : AB :: sin PAB : PB
or sin 2½ : 15 :: sin 10 : PB
and PB = 15 . sin 10 . cosec 2½
log 15 = 1·176091
10 sin = 9·965615
2½ cosec = 10·326613
PB 29·4 = 1·468319

The distances at either observation were 20·18, and 29·4 miles.

PART II.

NAUTICAL ASTRONOMY.

NAUTICAL ASTRONOMY.

INTRODUCTION.

THIS, the more extensive part of our subject, commonly known as Nautical Astronomy, or more accurately "Cœlo-Navigation," implies a knowledge of the various methods by which the zenith of a place is determined from observations of the heavenly bodies.

We have said that the earth is a body of spherical form, which revolves on its axis once in twenty-four hours; thus, the heavenly bodies will appear in succession above the horizon of any place, rising from the eastern portion and setting in the western part.

In Nautical Astronomy, however, the reverse movement is supposed to take place, The earth is conceived to be the centre of a sphere of indefinite magnitude, which revolves on an axis coincident in direction with that of our planet, which is presumed to remain at rest. The sun, moon, and other heavenly bodies are regarded as points on this extended sphere, moving over its surface at equal distances from the earth's centre, and in a contrary direction to that in which the earth itself actually revolves, —*i.e.*, from east to west.

By the revolution of the earth round the sun, that heavenly body appears to travel through certain constellations in the same time as that in which the earth accomplishes her orbit round him ; this is regarded in Nautical Astronomy as a real motion of the sun, the earth remaining at rest.

We now proceed to introduce those definitions upon which the subject is based.

CIRCLES.

Fig 11.

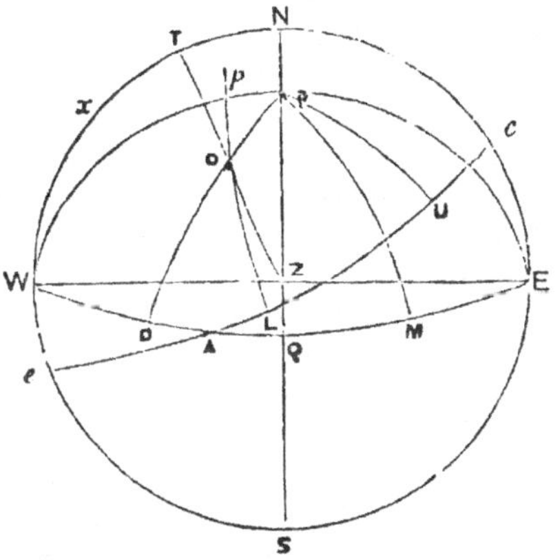

In the preceding figure, which represents the projection of the celestial sphere on the plane of the horizon, the observer is supposed to be placed at the centre of that circle, WNES, or immediately under the zenith, Z, which is a point on the celestial meridian, NZS.

It is evident that if the plane of the observer's meridian on the earth be produced till it meets the indefinite sphere surrounding the earth, which is called the **Celestial Concave**, a similar circle will be marked out on that sphere ; this is known as the **Celestial Meridian.** The point vertically over the observer's place is called his **Zenith**, and the corresponding point in the opposite hemisphere his **Nadir.** The meridian in the figure is NS.

Again, if the plane on which the observer stands (the centre of which is the point which marks his position) be produced to meet the celestial concave, we shall have another great circle called the **Sensible Horizon**, WNES. Similarly, an imaginary plane parallel to this, and passing through the earth's centre, will, if produced, mark on the celestial sphere a great circle called the **Rational Horizon.**

The axis of the earth produced both ways to meet the celestial concave forms the **axis of the Heavens,** and the two opposite points where it meets the celestial sphere, are called the **Poles of the Heavens.**

If the plane of the earth's Equator be similarly produced, we have the **Celestial Equator or Equinoctial,** WQE.

The Prime Vertical, WZE, is that great circle of the heavens which passes through the Zenith and Nadir, and the east and west points of the Horizon.

The six o'clock hour circle, WPE, is a great circle passing through the Poles of the Heavens and the east and west points of the Horizon.

Circles of altitude, (*e.g.* ZOT), are those great circles which pass through the Zenith and Nadir and cut the horizon at right angles. They are also called **Vertical circles**, or **Azimuth circles.**

Circles of declination, (*e.g.* POD), are those great circles which pass through the Poles of the Heavens and cut the Equator at right angles. They are also called **Meridians** and **Hour Circles.**

The circle which the sun appears to describe in a year amongst the fixed stars, by the revolution of the earth about him, is called the **Ecliptic** (*eAc*). The sun would seem to pass, in that interval, through twelve groups of stars, or constellations, called **signs of the Zodiac**, traversing each in a month. These are as follows:—

Aries (the Ram)	Leo (Lion)	Sagittarius (Archer)
Taurus (the Bull)	Virgo (Virgin)	Capricornus (Goat)
Gemini (the Twins)	Libra (Balances)	Aquarius (Water Bearer)
Cancer (the Crab)	Scorpio (Scorpion)	Pisces (Fishes)

The sun enters the Constellation Aries, on or about 21st March ; Cancer 21st June ; Libra, 21st September, and Capricorn, 21st December. In each of the signs some convenient point is selected ; *e.g.*, the first point of Aries, &c., for the purpose of estimating certain measurements. The first points of Aries and Libra are called respectively, the **Vernal** and **Autumnal Equinox**, and those of Cancer and Capricorn, the **summer** and **winter Solstice.**

The **Equinoctial points**, Aries, (A) and Libra, are those, in which the Ecliptic and the Celestial Equator appear to cut one another. The **Solsticial points** are situated on the Ecliptic, 90° from the Equinoctial points.

The **Colures** are those two meridians which pass through the Equinoctial and Solsticial points.

Great circles perpendicular to the ecliptic and intersecting one another in two opposite points called the Poles of the Ecliptic are called **circles of Celestial latitude.** The **Poles of the Ecliptic** are situated on the surface of the Celestial sphere 90° everywhere from the ecliptic.

Small circles parallel to the horizon are called **parallels of altitude**, and that parallel which is situated 18° beneath the horizon, is called the **Twilight Circle.**

Similarly, small circles parallel to the Equator are called **Parallels of declination**.

The sun and moon together exerting an attraction upon the protuberant part of the earth at the equator, the first point of Aries travels slowly westward in the Ecliptic, at the rate of 50″ a year ; this motion is named the **Precession of the Equinoxes.**

MEASUREMENTS.

The circles of the heavens above named being thoroughly understood, it will be now necessary to commit to memory, and to recognise in the figure, the following measurements in common use in Nautical Astronomy.

The **apparent altitude** of a heavenly body is that arc or portion of a circle of altitude intercepted between the body and the sensible horizon. In the figure OT measures the altitude of O. The complement of the altitude is the **Zenith Distance**, ZO, which is an arc of the same circle between the object and the zenith.

The **true altitude** of a heavenly body is the arc of a circle of altitude between the body and the rational horizon, and is, in fact, the altitude that object would have if observed from the earth's centre.

Declination of a body is the arc of a circle of declination between the object and the Equator, viz:—OD. The complement of the declination is the **Polar Distance**, PO, which is sometimes also found by adding 90° to the declination as hereafter explained ; or, the arc of the circle of declination between the body and the pole.

Latitude is that arc of a circle of latitude between the object and the Ecliptic, viz., OL.

Longitude is the arc of the Ecliptic between the first point of Aries, and the circle of latitude over the object, viz,, AL.

Amplitude is the arc of the horizon between the east point and the body rising, or the west and the body setting. If, in the figure, an object rose or set at *x*, W*x* would be its amplitude.

Azimuth is an angle at the zenith between the meridian of the observer and the circle of altitude over the body, viz., PZO.

The Hour Angle or Meridian Distance of a body is the angle at the pole between the meridian of the observer, and a circle of declination over the body, viz., OPZ.

Right Ascension is an angle at the pole, or the corresponding arc of the Equator, between the meridian of the first point of Aries and meridian over the body. This is estimated easterly in the order of the signs of the Zodiac, and is represented in the figure by the arc AD, or the angle APD.

TIME MEASURES.

In common phraseology the sun marks out the days and hours by its passage in the heavens, the interval between its rising and setting constituting a common day. This method of measurement by the **true or apparent sun**, is necessarily inexact from either or both of two causes, viz., (1) The motion of the sun in the Ecliptic is irregular, the spaces passed through by him in a certain interval being unequal ; and (2) his motion is in the Ecliptic, and not in the equator.

Astronomers, therefore, have recourse to an imaginary body called the **mean sun**, which is conceived to travel at an uniform rate upon the Equator, and therefore marks out regular intervals of time.

The hypothesis may be stated thus:—

(1.) An imaginary star is supposed to set out with the sun upon the Ecliptic from that point where he is nearest the earth, and his motion is most rapid, and to travel at an uniform pace on the Ecliptic. The sun and star will alternately be in advance of each other, and they will arrive at the opposite point of the Ecliptic whence they set out. This continuing round the Ecliptic and the motion of the star being substituted for that of the sun, the inequality of the sun's motion is compensated for.

(2.) When the star, in its course, reaches the first point of Aries, let another imaginary body called the **mean sun** set out with him, and travel along the equator at the rate of the star in the Ecliptic. They will meet together at Libra, each having traversed 180° at the same rate. If now the transits of the mean sun be substituted for those of the star the two causes of inequality in the solar days are compensated.

This gives rise to the **Equation of Time**, an angle at the pole, or corresponding arc of the equator, between the meridian over the mean sun and that over the true sun, viz., MPU in the figure, by which mean time is reduced to apparent time, and *vice versa*. The Equation of Time consists of two parts, arising from the two causes above stated, viz., the unequal motion of the sun in the Ecliptic, and the fact of its moving in that circle instead of in the equator. The latter part vanishes at or near the period of the Equinoxes, viz., March and September, and the other near the times of the Solstices, viz., June and December. The equation of time, therefore, never entirely vanishes, and never exceeds 16^m.

Since now, an uniform measure of time is established, we may proceed with the following additional definitions, viz:—

Apparent noon is the instant when the apparent or true sun comes on the meridian of a place, and **mean noon**, when the mean sun reaches that point, either marking the commencement of an apparent or mean solar day.

An **apparent Solar Day** is the interval between two successive transits or passages of the apparent sun over some fixed point of the heavens ; and a **Mean Solar Day** the corresponding interval in the case of the mean sun.

The interval of two successive transits of the first point of Aries over the same point of the celestial sphere, is termed **a sidereal day** which is 4^m shorter (nearly) than a mean solar day.

If we conceive a clock, marked round to 24 hours, to be placed at the pole, the 24 being placed on the meridian of the observer, and the sun in its revolution round the celestial sphere, to carry a meridian and mark the time on the face of this dial, we form an idea of the method of estimating an **Astronomical day.** That day is said to commence at noon, and to end on the following noon, and is reckoned round to 24 hours, in contradistinction to the **Civil** day which begins at 12 midnight. The 24 hours of the one are marked by the 12 hours a.m., and 12 hours p.m. of the other. These letters have no place in the problems of Nautical Astronomy, all **Civil Time** being reduced previously to **Astronomical Time.**

Fig 12.

Fig A. Fig B.

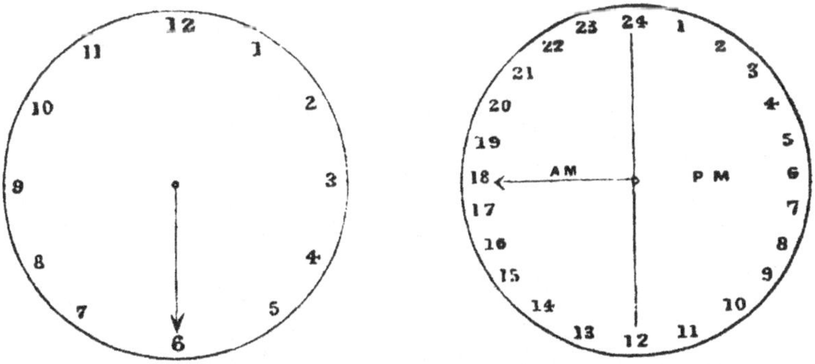

Let Fig A represent the face of an ordinary clock, and Fig B that of an astronomical clock. It is clear that the pointer of A makes two revolutios, while that of B makes but one. Let the two clocks mark the same time, and when the pointer of A is at 12, that of B will be at 24; this is the commencement of an astronomical day. The pointer of A will then travel from 12 to 12, while that of B goes from 24 to 12. These 12 hours of A are P.M. and are marked out on B as such, while the remaining 12 are A.M.

Now it is evident that, when A shows 6 hours, from noon, B will show 6; but, when A shows the next 6 hours, B will show 18 hours. Any number of hours, then, marked out by A is ambiguous, unless designated by A.M. or P.M., while the necessity of these terms for B does not exist; thus, 6 A.M. is simply 18 hours; 9 A.M. is 21 hours, &c., reckoning of course from the preceding noon when the pointer last stood at 24^h or 0^h. This will be made use of in every subsequent problem. Therefore—

Apparent Time is the **Westerly** hour angle of the apparent or **True** sun.

Mean Time is the **Westerly** hour angle of the Mean sun.

Sidereal Time is the **Westerly** hour angle of the first point of Aries.

These definitions are most important, and must be carefully borne in mind, in order to prevent confusion hereafter.

The Right Ascension of the Meridian is the angle at the pole or corresponding arc of the Equator, beween the Meridian over the first point of

Aries, and the Meridian itself; this expression is synonymous with Sidereal time.

The **Obliquity of the Ecliptic** is the angle at which the Ecliptic and Equator cut one another, the exact amount of which is 23° 27′ 45″, the inclination of the plane of the earth's Equator to the plane of its Orbit.

Projection of the Celestial Sphere on the plane of the Observer's Meridian.

Fig 13.

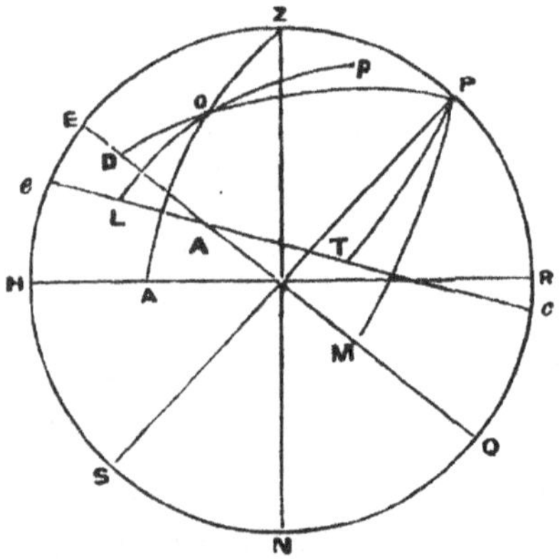

The corresponding points indicated in the other projection are here seen from a different position. HZRN is the celestial meridian; EQ the Equator will appear as a straight line, as will also ZN and PS the Prime Vertical, and six o'clock hour circles: *ec* represents the plane of the Ecliptic; ZOA a circle of altitude; O the celestial object; POD, and *po*L circles of declination and latitude respectively. M the place of the Mean Sun on the Equator, and T that of the true Sun in the Ecliptic, then—

OA	is the	altitude of O	OL	is the	latitude of O
ZO	,,	Zenith dist of O	AD	,,	right ascension of O
OD	,,	declination of O	AL	,,	longitude of O
PO	,,	polar distance of O			

MPT represents the equation of time; A the first point of Aries; EA*e* the obliquity of the Ecliptic = 23° 27′ 45″; *p* the pole of the Ecliptic, AM is the Right Ascension of the Mean sun, and AE that of the Meridian; PZO is the Azimuth of O; OP*z* its Hour Angle; PR the Latitude of the Place of Observation; and P*z* its co-Latitude.

The occultation of a body takes place when another passes over its disc.

The **Heliocentric place** of a heavenly body, is that which it would have if seen from the centre of the sun; the **Geocentric**, from the centre of the earth.

Dip of the Horizon is the angle at the observer's eye; between a horizontal line and a tangent to the earth's surface drawn from that point.

Fig 14.

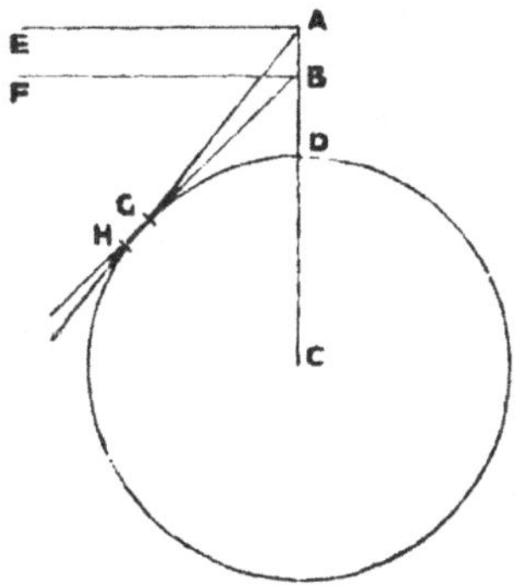

Suppose an observer to be elevated at A above the surface of the earth. Then AE is a horizontal line, and AH a tangent to the earth's surface from that point. The angle EAG is the dip of the **Apparent Horizon** AH.

If he were placed at B, the angle FBG which would then be the dip, would be smaller than EAH.

We conclude, therefore, that the greater the observer's elevation, the larger is the dip and *vice versa.*

Now an altitude of a heavenly body measured down to the horizon AH, must be larger than if it were measured to BG; hence dip increases an altitude, and is, therefore, **subtractive** in every case, as will be seen subsequently.

It is clear that when the point B travels to D, or when the observer is situated on the sea surface, the line BG coincides with the sea horizon and the dip vanishes.

Refraction is the angle at the observer's eye between a line drawn to the apparent place of a body and another to its true place.

Fig 15.

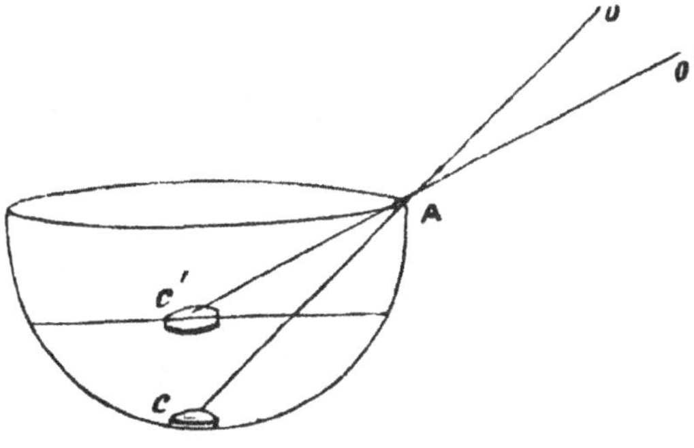

To illustrate: let a coin *c* be placed in an empty basin when an observer at O can just see it over the edge of the basin. Let the vessel be now partly filled with water, the coin will become visible at O', appearing to have been raised from *c* to *c'*.

The ray of light, in this case, passes from a denser to a rarer medium, but this yet illustrates refraction. The reverse is the case in the atmosphere, which surrounds the earth as an envelope, growing denser and denser as we approach the surface ; here the rays pass from a rarer to a denser medium.

Fig 16.

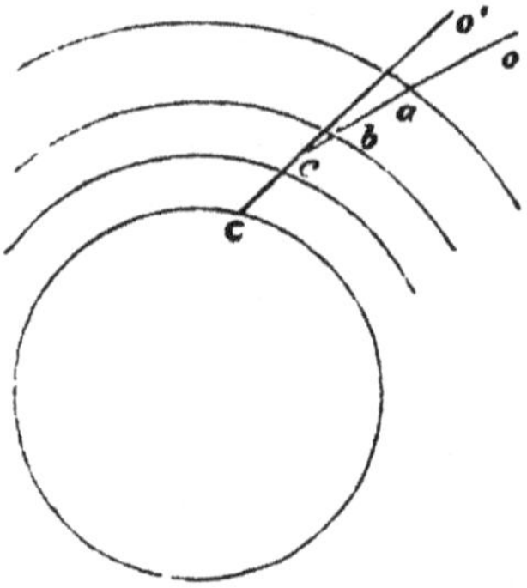

A ray of light from a body O will be successively bent as it passes through the atmosphere in the directions *ab*, *bc*, &c., till, when it reaches the observer at C, the body will appear to him to be in the direction C*c*, of the last ray of light or at O′. Hence as refraction tends to increase an altitude, it is, like the dip, a subtractive quantity.

Parallax is the angle at the centre of a heavenly body subtended by the earth's radius.

Fig. 17.

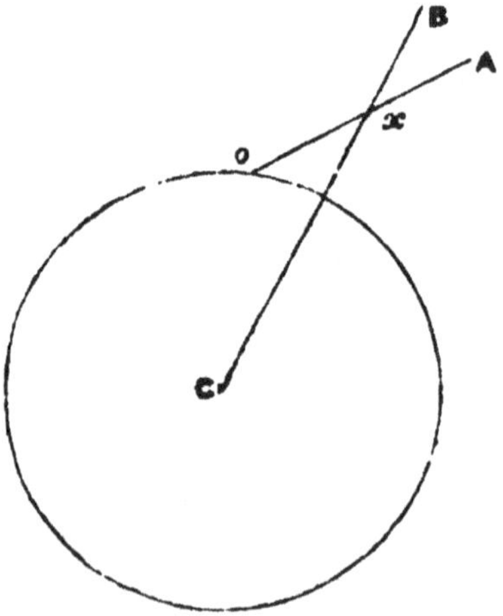

In the accompanying figure a heavenly body *x*, as seen from O, would appear in the direction A : but if seen from C, would appear at B. The angle O*x*C or B*x*A is the parallax, and as altitudes are taken from the surface of the earth O, heavenly bodies appear lower than if seen from the earth's centre. As parallax then tends to decrease an altitude, it is an additive quantity.

In observations of the sun, moon, or a planet, the altitude of either the upper or lower limb of the object is taken. In order, therefore, to obtain the altitude of the object's centre, which is the quantity sought, it is evident that the altitude of the upper limb is too great for that of the centre by half the diameter, or the semi-diameter, and the altitude of the lower limb less than that of the centre by the same amount. Hence in lower limb observations the SD is added, and in upper limb subtracted to obtain the altitude of the body's centre.

We defer the consideration of Index Error of the sextant, which is another correction to be applied to an altitude, until that instrument is treated of subsequently.

SUBORDINATE COMPUTATIONS.

I. Conversion of Time and Arc.

By the revolution of the earth upon its axis, 15° of its surface, measured on the Equator, pass under the sun in one hour of time; or, more familiarly, the sun travels from east to west over that portion of the Equator in that interval. As longitude is measured on the Equator, we have at once the connection between it and time: *i.e.*,

$$15^\circ = 1 \text{ hour.}$$
$$15' = 1 \text{ min.}$$
$$15'' = 1 \text{ sec.}$$

And as the motion of the sun is from east to west, apparently, it follows that all places east of a certain point would have later time, and all west of it earlier time. All time is referred by Englishmen to the meridian of Greenwich, and the elements in the Nautical Almanac are computed only for Greenwich time. The difference of time, therefore, between Greenwich and any place is the longitude of that place, so that to reduce time at place to that at Greenwich the longitude must be reduced to time and applied to the former.

Example.—Convert 17° 39′ 42″ into time.

Hence we must divide by 15, for 15° = 1 hour, &c.; or which will be found more convenient, we may multiply by 4, and divide by 60.

$$\begin{array}{r} 17^\circ \;\; 39' \;\; 42'' \\ 4 \\ \hline 6{,}0)7{,}0 \;\;\; 38 \;\;\; 48 \\ \hline 1^h \;\; 10^m \;\; 38^s \tfrac{48}{60} \\ = 1^h \;\; 10^m \;\; 38^s.8 \end{array}$$

Example.—How much arc corresponds with $7^h\ 39^m\ 47^s\cdot6$?

$$
\begin{array}{rrr}
7^h\ 39^m & 47 & \cdot 6 \\
60 & & \\
\hline
4)459 & 47 & 36 \\
114^\circ & 56' & 54'' \\
\hline
\end{array}
$$

EXAMPLES FOR EXERCISE.

		°	′	″			h	m	s
1.	Convert	139	47	47	into time.	Ans.	9	19	11·13
2.	,,	47	20	30	,, ,,	,,	3	9	22
3.	,,	122	29	47	,, ,,	,,	8	9	59·13
4.	,,	12	15	18	,, ,,	,,	0	49	1·2
5.	,,	79	16	45	,, ,,	,,	5	17	7
6.	,,	32	12	44	,, ,,	,,	2	8	50·93
		h	m	s			°	′	″
7.	,,	7	42	47·5	into arc.	,,	115	41	52
8.	,,	10	39	10	,,	,,	159	47	30
9.	,,	4	49	37	,,	,,	72	24	15
10.	,,	2	37	20	,,	,,	39	20	0
11.	,,	8	31	12	,,	,,	127	48	0
12.	,,	3	42	40	,,	,,	55	40	0

II. To Find the Corresponding Greenwich Date from Date at Place.

(*a*) Convert the civil date at place into the corresponding astronomical date.

(*b*) Apply the longitude in time; if east, subtract; if west, add.

Example.—January 12th, at $9^h\ 30^m\ 40^s$ A.M., longitude 47° 18′ 45″ W; find the Greenwich date.

Date.

$$
\begin{array}{rrrr}
11^d & 21^h & 30^m & 40^s \\
 & 3 & 9 & 15 \\
\hline
12 & 0 & 39 & 55 \\
\hline
\end{array}
$$

Longitude.

$$
\begin{array}{rrr}
47^\circ & 18' & 45''\ \mathrm{W} \\
 & & 4 \\
\hline
6{,}0)18{,}9 & 15 & 0 \\
\hline
3 & 9 & 15 \\
\hline
\end{array}
$$

Example.—March 12th, at $2^h\ 30^m\ 45^s$ P.M., in longitude 79° 30′ 40″ E; find the Greenwich date.

Date.

$$
\begin{array}{rrrr}
12^d & 2^h & 30^m & 45^s \\
 & 5 & 18 & 2\cdot7 \\
\hline
11 & 21 & 12 & 42\cdot3 \\
\hline
\end{array}
$$

Longitude.

$$
\begin{array}{rrr}
79^\circ & 30' & 40''\ \mathrm{E} \\
 & & 4 \\
\hline
6{,}0)31{,}8 & 2 & 40 \\
\hline
5^h & 18^m & 2^s\cdot7 \\
\hline
\end{array}
$$

EXAMPLES FOR EXERCISE.

1. March 28th, at $4^h\ 30^m\ 40^s$ A.M., longitude 62° 30′ E..
Ans. $27^d\ 12^h\ 20^m\ 40^s$.

2. July 12th, at $7^h\ 31^m\ 20^s$ P.M., longitude 72° 10′ W.
Ans. $12^d\ 12^h\ 20^m\ 0^s$.

3. September 17th, at $3^h\ 44^m\ 10^s$ A.M., longitude 42° 20′ E.
Ans. $16^d\ 12^h\ 54^m\ 50^s$.

III. Correction of Elements in the Nautical Almanac.

The pages of each month are distinguished by Roman capitals in the corner, marked I., II., &c.

Page I. contains the sun's apparent right ascension, his declination with the change in one hour; the sidereal time of the semi-diam. passing the meridian, and the equation of time with the change in one hour, calculated for Greenwich apparent time.

Page II. contains nearly the same elements, but calculated for Greenwich mean time, together with the sun's semi-diam. and the sidereal time.

Pages III and IV of the month are occupied chiefly with the moon's elements; her semi-diam. and horizontal parallax being tabulated on page III for noon and midnight, and her meridian passage at her upper and lower transits on page IV, which are the elements we shall have occasion to use.

From page V—XII are contained the moon's right ascension and declination with the average change in 10 minutes, calculated for each hour of the day.

Pages XIII—XVIII contain tables of lunar distances calculated for every three hours of mean time.

The right ascensions and declinations of the principal stars and planets are given farther on, and towards the end is a table for converting intervals of solar time into equivalent intervals of sidereal time.

Example.

(A) To Correct the Sun's Declination for a given Greenwich Date.

Note.—If the time at place be apparent time take from page I. of the almanac, but from page II. if mean time. The change column for both is on page I.

If the sun's declination be increasing add the correction, but subtract if decreasing.

Multiply the change by the hours and minutes of the Greenwich date.

1. Find the sun's declination for January 12th, 1876, at $2^h\ 30^m\ 45^s$ P.M., mean time, in longitude 39° 20′ E.

Greenwich Date.	Longitude.
$12^d\ 2^h\ 30^m\ 45^s$	39° 20′ E
2 37 20	4
11 23 53 25	6,0)15,7 20
	2 37 20

Declination (page 2.)	Correction.	
S 21° 51′ 58″·1	Change in 1^h =	23″·14 —
—9 13 ·0	Hours past noon	23·9
21 42 45 ·1		20826
		6942
		4628
		6,0)55,3,046
		—9′ 13″

2. Find the sun's declination for March 19th, 1876, at $10^h\ 45^m\ 27^s$ P.M., apparent time long 120° 30′ W.

Greenwich Date.	Longitude.
$19^d\ 10^h\ 45^m\ 27^s$	120° 30′ W
8 2 0	4
19 18 47 27	6,0)48,2 0
	$8^h\ 2^m\ 0^s$

Declination (Page 1.)	Correction.
S 0° 17′ 49″·3	—59″·27
—18 34 ·3	18·8
N 0 0 45 ·0	47416
	47416
	5927
	6,0)1114,276
	18′ 34″·3

In this example, the sun crosses the equator from S to N in the interval, and the change being greater than the declination, which is decreasing from S to N, the latter is subtracted from the former and the name reversed. In all similar examples which may occur, care must be taken to note this.

(B) To find the Equation of Time for a given Greenwich Date.

July 12th, 1876, at 2^h 40^m A.M., mean time, in long 73° 20′ W; find the equation of time.

Greenwich Date.	Longitude.
11^d 14^h 40^m 0^s	73° 20′ W
4 53 20	6,0)29,3 20
11 19 33 20	4 53 20

Eq. of Time.	Change.
5^m 21^s·88	+^{s}·305
5 ·97	19·6
5 27 ·85	1830
	2745
	305
	5·9780

(C) To find the Mean Sun's Right Ascension for a given Date.

Note.—Take out the sidereal time, (page II) for the given day; then apply the **acceleration**, from the table which converts solar into sidereal time for the hours, minutes and seconds of the Greenwich date.

Example.—Find the mean sun's right ascension for July 12^d 18^h 37^m 40^s, Greenwich Date, 1876.

Sidereal Time (page II) for 12th	=	7^h 22^m 53^s · 50
Acceleration, page 496, for	18^h =	2 57 ·41
	37^m =	6 ·08
	40^s =	·11
Right ascension of mean sun	=	7 25 57 ·10

(D) To find the Moon's Semi-diameter and Horizontal Parallax for a given Date.

Note.—(*a*) If the Greenwich date be less than 12^h, take from the column for noon; if greater, for midnight.

(*b*) To find the change; the difference between noon and midnight of same day, if taken for noon; between midnight and following noon if taken for midnight.

(*c*) Multiply the change by the hours past noon or past midnight, and add or subtract as the elements increase or decrease.

Example.—Required the moon's semi-diameter and horizontal parallax for January 8^d 6^h 30^m, Greenwich date.

Semi-diam (noon on 8th).	Change.	
16′ 34″ · 2	(Noon and midnight)	+ 3″·4
+1 · 9	Hours past noon	6 · 5
16 36 · 1		170
		204
		12)22·10
		1·9

Hor¹ parallax (noon on 8th).	Change.	
60′ 42″ · 6	(Noon and midnight)	+12″·3
+6 · 6	Hours past noon	6 · 5
60 49 · 2		615
		738
		12)79·95
		6·6

N.B.—The change is that in 12^h; therefore dividing by 12 gives the change in 1^h, and multiplying by 6.5 gives the change in 6^h 30^m of Greenwich date.

Example.—Required the moon's semi-diameter and horizontal parallax for July 12^d 17^h 42^m, Greenwich date.

Semi-diam (midnight on 12th).	Change.	
15′ 21″ · 4	(Midnight and noon following)	+5″·9
+2 · 8	Hours past midnight	5 · 7
15 24 · 2		413
		295
		12)33·63
		2·8

Hor¹ parallax (midnight on 12th).	Change.	
56′ 15″ · 8	(Midnight and noon following)	+21″·6
+10 · 2	Hours past midnight	5 · 7
56 26 · 0		1512
		1080
		12)123·12
		10·2

(E) To find the Greenwich Date of the Moon's Meridian Passage.

Note.—(*a*) If on the given day the passage (upper transit) be greater than 12^h, take for preceding day; if less take for the same day.

(*b*) To obtain the moon's retardation; if the longitude be east, take the difference between that time of passage and the previous one; if west, between that and the following.

(*c*) Multiply by the hours and minutes of the longitude in time and divide by 24. Add if the long be west, subtract if east.

(*d*) Apply the longitude in time to the meridian passage.

Example.—July 24th, 1876, in long 72° 30′ W, find the Greenwich date of the moon's meridian passage.

Meridian Passage.

	d	h	m
On 24th	24	3	9·2
Retardation		+	9·2
	24	3	18·4
Long	+	4	50·0
Green^h date	24	8	8·4

Retardation.

	m
Between 24th—25th	+46.0
	4.8
	3680
	1840
	4)220·80
	6)55·20
	+9·2

Longitude.

	°	′
	72	30 W
		4
6,0)29,0		0
	4	50·0

Example.—July 15th, 1876, in long 132° 40′ E.

Meridian Passage.

	d	h	m
On 14th	14	18	19·1
Retardation		−	17·1
	14	18	2·0
Long		8	50·7
Green^h date	14	9	11·3

Retardation.

	m
Between 14th—13th	−46.8
	8.8
	3744
	3744
	4)411·84
	6)102·9
	17·1

Longitude.

	°	′
	132	40 E
6,0)53,0		40
	8	50 40

(F) To find the Moon's Right Ascension and Declination for a given date.

Note.—(*a*) Take each out for the given day and hour.

(*b*) Divide the change in 10^m by 10 (or shift the decimal point one place to the left); multiply the change by the minutes and seconds of the Greenwich date. Add if increasing; subtract if decreasing.

Example.—Required the moon's right ascension and declination for January 18[d] 3[h] 40[m] 42[s], Greenwich date.

	RA	Change	Dec	Change
	h m s	s	° ′ ″	″
For 18th 3[h] =	13 51 7·73	+1·90	S 14 6 23·7	+12·93
	1 17·33	40·7	8 46·2	40·7
	13 52 25·06	1330	14 15 9.9	9051
		760		5172
		6,0)77·330		6,0)52,6·251
		1[m] 17[s]·33		8·46·2

(G) TO FIND THE TRUE GREENWICH DATE FROM A LUNAR DISTANCE.

Note.—(*a*) Take out the distance between the moon and object for the next less number of hours to those of the approximate Greenwich date (pages xiii—xviii Nautical Almanac), with the accompanying proportional logarithm.

(*b*) Take the difference of the distances; find the proportional log of this difference from the table of proportional logarithms.

(*c*) Take the difference between the proportional logs, and from the same table take the number of hours, minutes, and seconds corresponding.

(*d*) Add the number of hours over the column in the Almanac, from which the distance was taken.

(*e*) Enter the table of second differences in the Nautical Almanac, with the difference of the proportional logarithms in 3[h] as given in the Almanac, and also the number of hours and minutes mentioned in (*c*). Add this correction if the proportional logarithms in the Almanac increase, subtract if they decrease.

Example.—December 18th, 1879, at about 4 P.M., Greenwich date nearly, the true distance between the sun and moon was 68° 51′ 0″; find the true Greenwich date

			Proportional logarithms.
Given distance =	68° 51′ 0″		
Almanac dist iii[h] =	68 21 3	. . .	2981
Difference	0 29 57	. . .	7789
Interval	0[h] 59[m] 30[s]	. .	4808
Add	3		
	18[d] 3 59 30		

CORRECTION FOR SECOND DIFFERENCES.

Interval 0[h] 59[m] 30[s]		gives 6[s] nearly to be added	See page 484 for N.A., 1879
Diff of prop[l] log[s] from iii to vi[h]	19 (increasing)		

Date as above	18[d] 3[h] 59[m] 30[s]
Correction	+ 6
True Green[h] date	18 3 59 36

(H) To find the Sun's Hour Angle from the Mean Time at Place.

The Hour Angle of the Apparent or True Sun is measured in Apparent Time; hence to the given Mean Time the Equation of Time must be applied. The Mean Time at place must be reduced to Astronomical Time.

Example.—Given the Mean Time $9^h\ 40^m\ 40^s$ A.M., Equation of Time $+\ 2^m\ 30^s{\cdot}7$; find the Sun's Hour Angle. The Equation Time is taken $\pm$ to mean Time as given on page ii of the Almanac.

Mean Time	21^h	40^m	40^s
Equation Time	+	2	30·7
	21	43	10·7
Subtract from	24	0	0
	2	16	49·3
	60		
	4)136	49	18
Hour Angle	34°	12′	19·5

Example.—Given the Mean Time $2^h\ 30^m\ 40^s$ P.M., Equation of Time $-\ 1^m\ 40^s.6$; find the Sun's Hour Angle.

Mean Time	2^h	30^m	40^s
Equation Time	—	1	40·6
	2	28	59·4
	60		
	4)148	59	24
Hour Angle	37°	14′	51″

(I) To find a Star's Hour Angle from the Mean Time at Place.

Example.—On March 10th, 1876, at $6^h\ 40^m\ 30^s$ P.M., Mean Time, the Right Ascension of Capella was $5^h\ 7^m\ 32^s{\cdot}5$, and the Mean Sun's Right Ascension was $23^h\ 14^m\ 0^s{\cdot}5$: find the Star's Hour Angle.

Mean Time	6^h	40^m	30^s
+ RA of Mean Sun	23	14	0·5
	29	54	30·5
– RA of Star	5	7	32·5
	24	46	58·0
Reject	24	0	0
	0	46	58
	60		
	4)46	58	0
Hour Angle	11°	44′	30″

Example.—At $5^h\ 10^m\ 14^s$ P.M., the RA of the mean sun was $23^h\ 45^m\ 32^s.9$, and the RA of Sirius was $6^h\ 39^m\ 40^s.8$: find the star's hour angle.

Mean Time	5^h	10^m	14^s
+RA of Mean Sun	23	45	32·9
	28	55	46·9
−RA of Star	6	39	40·8
	22	16	6·1
Subtract from	24	0	0
	1	43	53·9
	60		
4)	103	53	54
Hour angle	25°	58′	28″

N.B.—This rule answers also for moon and planet.

(J) To Correct an Observed Altitude.

Example.—January 28th, 1876, given the observed altitude of the sun's lower limb, (⨀̲) 29° 42′ 10″, Index Error + 1′ 15″: height of eye 20 feet; find the true alt of the centre.

Obs alt ⨀̲	29°	42′	10″	
IE +		1	15	
	29	43	25	
(Height of eye 20 ft) dip	−	4	24	(from table)
	29	39	1	
Refn.		−1	40	(from table)
	29	37	21	
Semi-diamr.		+16	18	(Nautl. Almc. page II)
	29	53	39	
Parallax	+		8	(from table)
True alt ⨀s. centre	29	53	47	

N.B.—The semi-diameter is **added** when **lower** limb is observed, but, **subtracted** when **upper** limb is taken. For other corrections see foregoing explanation.

Example.—Given the observed altitude of a star 39° 47′ 20″, Index Error + 2′ 30″; height of the eye 24 feet: find its true altitude.

Obs alt	39°	47′	20″
IE +		2	30
	39	49	50
Dip −		4	49
	39	45	1
Refn.	−	1	9
Stars true alt:	39	43	52

N.B.—The star's altitude is unaffected by either semi-diameter or parallax. The distances of the fixed stars from the earth are too great to afford any sensible parallax.

SUBORDINATE COMPUTATIONS.

Examples for Exercise.

1876. (A) (B)

1. Find the sun's declination on January 8th, $9^h\ 42^m$ Greenwich date.
Ans. S 22° 14′ 33″.

2. Correct the sun's declination and the equation of time for March 12th, in long 82° 30′ E, at $10^h\ 42^m\ 30^s$ P.M., mean time at place.
Ans. Dec S 2° 58′ 29″
Eq T $9^m\ 43^s.0$

3. Correct the sun's declination for September, $22^d\ 21^h\ 40^m$ Greenwich date.
Ans. S 0° 16′ 11″.6

(C)

4. Find the mean sun's right ascension for April, $12^d\ 7^h\ 30^m\ 44^s$, Greenwich date.
Ans. $1^h\ 25^m\ 20^s.7$

5. Also on May 10th, in long 79° 30′ W, at $2^h\ 30^m$ A.M.
Ans. $3^h\ 13^m\ 48^s.9$

(D)

6. Find the moon's semi-diameter for April $21^d\ 5^h\ 30^m$, Greenwich date
Ans. 15′ 46″.4

7. Also for July $18^d\ 18^h\ 12^m$, Greenwich date.
Ans. 16′ 38″.6

8. Find the moon's horizontal parallax for December, $10^d\ 19^h\ 44^m$, Greenwich Date.
Ans. 56′ 19″.4

(E)

9. January 21st, 1876, in long 81° 30′ W: find the Greenwich date of the moon's meridian passage.
Ans. $21^d\ 14^h\ 2^m.7$

10. January 6th, in long 84° 30′ E, find the Greenwich date of the moon's meridian passage.
Ans. $5^d\ 13^h\ 20^m.5$.

(F)

11. Find the moon's right ascension and declination for January $12^d\ 17^h\ 30^m\ 42^s$ Greenwich date.

Ans. $9^d\ 32^h\ 15^m$ and N 18° 10′ 30″.

12. Also for March $21^d\ 17^h\ 20^m\ 48^s$ Greenwich date.

Ans. $21^h\ 27^m\ 6^s.1$ and S 18° 40′ 15″.

(G)

13. On February 16th, 1876, at $8^h\ 30^m$ A.M., mean time nearly, in longitude 175° E, the true distance between the sun and moon was 104° 47′ 16″; find the true Greenwich date.

Ans. February $15^d\ 8^h\ 47^m\ 34^s$.

14. October 30th, 1876, at $8^h\ 20^m$ P.M., mean time nearly, in longitude 2° 30′ W, the true distance between Saturn and the moon was 39° 40′ 56″; find the true Greenwich date.

Ans. October $30^d\ 8^h\ 39^m\ 54^s$.

(H)

15. Given the mean time at place, $2^h\ 30^m\ 30^s$ P.M., Equation of time $2^m\ 30^s\cdot5$ (to be subtracted); find the true sun's hour angle.

Ans. 36° 59′ 52″.

16. Given the true apparent time $9^h\ 40^m\ 50^s$ A.M.; find the sun's hour angle.

Ans. 34° 47′ 30″.

(I)

17. Given the mean time $9^h\ 30^m\ 30^s$ P.M., the sidereal time $2^h\ 40^m\ 10^s$, and a star's R.A. $10^h\ 1^m\ 42^s$; find its hour angle.

Ans. 32° 14′ 30″.

18. Given the mean time $5^h\ 40^m\ 40^s$ A.M., the sidereal time $13^h\ 21^m\ 40^s$, and star's R.A. $15^h\ 42^m\ 10^s$; find its hour angle.

Ans. 129° 57′ 30″.

(J)

19. Given the obs alt $\underline{\odot}$ 51° 27′ 10″, IE—2′ 0″; height of the eye 20 feet; find the true alt, the semi-diameter being 16′ 18″.

Ans. 51° 36′ 23″.

20. Given the obs alt of a star 59° 28′ 40″, IE + 1′ 15″; height of the eye 24 feet; find its true altitude.

Ans. 59° 24′ 32″.

21. Given the obs alt $\overline{\overline{\odot}}$ 150° 25′ 10″ IE + 1′ 14″; semi-diameter 15′ 47″; find the true altitude of its centre.

Ans. 74° 57′ 12″.

PRACTICAL METHODS OF NAUTICAL ASTRONOMY.

LATITUDE.

1. To find the Latitude from a Meridian Altitude of the Sun.

Note.—(*a*) Find the apparent Greenwich date of the sun's transit. (The time of the apparent sun's transit at any place is always $0^h\ 0^m\ 0^s$, for it marks the commencement of an apparent solar day).

(*b*) Correct the sun's declination.

(*c*) Correct the altitude, and subtract it from 90° which gives the Zenith distance, and name it.

(*d*) Apply the corrected declination, adding if the Zenith distance and declination are alike, and subtracting if unlike.

Example.

On January 10th, 1876, in long 84° 30′ W, the obs mer alt ☉ was 62° 30′ 10″, IE—2′ 10″: Zenith North (ZN), height of the eye 24 feet; find the latitude.

Greenwich Date.

$10^d\ 0^h\ 0^m\ 0^s$
+5 38 0
10 5 38 0

Longitude.

84° 30′ W
4
6,0)33,8 0
5 38 0

Declination (page I.)

S 22° 0′ 57″.7
— 2 3 .6
21 58 54 .1

Correction.

—22″.08
5 . 6
13248
11040
6,0)12,3.648
2′ 3″.6

Corrected Altitude.

Obs alt ⊙	62°	30′	10″	
IE	—	2	10	
	62	28	0	
Dip	—	4	49	
	62	23	11	
Ref	—		30	
	62	22	41	
SD	+	16	18	
	62	38	59	
Par^x^	+		4	
True alt	62	39	3	
Subtract from	90	0	0	
Zen dist	27	20	57	N
Dec^n^	21	58	54	S
Latitude	5	22	3	N

This is the most common method in use at sea of finding the latitude. The altitude of the sun is observed when **nearly** noon at the place, *i.e.*, when the sun is near the meridian, and continuous observations are taken till the altitudes are found to decrease. The **greatest altitude** of the set is the meridian altitude.

2. To find the Latitude from a Meridian Altitude of a Star.

Note.—(*a*) Correct the altitude and find the zenith distance.
(*b*) Apply the star's declination as by above rule.

Example.—January 12th, 1876, the mer alt of Algenib was 51° 30′ 10″ (ZS), IE — 2′ 10″; height of eye 14 feet; find the latitude.

Obs alt	51°	30′	10″	
IE	—	2	10	
	51	28	0	
Dip	—	3	41	
	51	24	19	
Ref	—		45	
True alt	51	23	34	
	90	0	0	
Zen dist	38	36	26	S
Dec	14	29	40	N
Latitude	24	6	46	S

3. To find the Latitude from a Meridian Altitude of the Sun below Pole.

Note.—(*a*) The sun comes to the meridian below pole 12 hours later than above pole, therefore the date at place of the observation will be 12ʰ 0ᵐ 0ˢ. Find the Greenwich date.

(*b*) Correct the declination and subtract it from 90°, which gives the polar distance.

(*c*) Correct the altitude.

(*d*) Add the polar distance to the true altitude, which gives the latitude required.

Example.—January 30th, 1876, in long 75° 20′ E, the obs mer alt ⨀ below pole was 5° 30′ 30″, index error + 50″; height of the eye 27 feet; find the latitude.

Greenwich date.

30ᵈ 12ʰ 0ᵐ 0ˢ	
5 1 20	
30 6 58 40	

Longitude.

75° 20′ E
4
6,0)30,1 20
5 1 20

Declination.

	S 17° 45′ 4″·1
	— 4 42 ·5
	17 40 21 ·6
	90 0 0
Polar dist	72 19 38 ·4

Correction.

—40″·94
6·9
36846
24564
6,0)28,2·486
4′ 42″·5

Corrected Altitude.

Obs alt ⨀	5° 30′ 30″
IE	+ 50
	5 31 20
Dip	— 5 7
	5 26 13
Ref	— 9 15
	5 16 58
SD	+ 16 16
	5 33 14
Parallax	+ 9
True alt	5 33 23
Polar dist	72 19 38
Latitude	77 53 1 S

N.B.—The latitude is of the same name as the declination, for the sun can only be seen upon the meridian below pole in a high latitude of that name, as will be found explained in the investigation of the rule.

4. To find the Latitude by a Meridian Altitude of a Star below Pole.

Note.—(*a*) Correct the altitude.

(*b*) Find the star's polar distance.

(*c*) Add the polar distance to the true altitude.

Example.—January 12th, 1876, the obs mer alt of β Hydri below pole was 3° 20′ 30″, IE — 2′ 10″, HE 20 feet; find the latitude.

Obs alt	3°	20′	30″
IE	—	2	10
	3	18	20
Dip	—	4	24
	3	13	56
Ref	—	13	39
True alt	3	0	17
Polar dist	12	2	29
Latitude	15	2	46 S

Star's dec	77°	57′	31″ S
Polar dist	12	2	29

EXAMPLES FOR EXERCISE.

1. January 10th, 1876, in long 84° 30′ W, the obs mer alt $\underline{\odot}$ was 62° 30′ 10″, (ZN) IE – 2′ 10″; height of the eye 24 feet: find the latitude
Ans. 5° 21′ 38″ N.

2. February 8th, in long 76° 20′ E, the obs mer alt $\overline{\odot}$ was 39° 21′ 20″, (ZS), IE + 1′ 15″; height of the eye 23 feet: find the latitude.
Ans. 66° 9′ 17″ S.

3. March 12th, in long 84° 30′ E, the obs mer alt $\underline{\odot}$ was 54° 20′ 30″, (ZN), IE—2′ 15″; height af the eye 21 feet: find the latitude.
Ans. 32° 21′ 37″ N.

4. April 14th, in long 110° 30′ W, the obs mer alt $\overline{\odot}$ was 62° 40′ 20″, (ZS), IE + 3′ 5″; height of the eye 22 feet: find the latitude.
Ans. 17° 52′ 46″ S.

5. May 5th, in long 32° 15′ E, the obs double mer alt $\underline{\underline{\odot}}$ (artificial horizon) was 102° 39′ 10″ (ZN), IE – 2′ 20″: find the latitude.
Ans. 54° 51′ 0″ N.

6. June 7th, in long 117° 40′ W, the obs mer alt $\underline{\odot}$ was 62° 17′ 30″ (ZS), IE + 1′ 20″; height of the eye 17 feet: find the latitude.
Ans. 4° 38′ 42″ S.

7. July 20th, in long 77° 30′ W, the obs mer alt $\overline{\odot}$ was 71° 20′ 30″, (ZN), IE — 1′ 15″; height of the eye 16 feet: find the latitude.

Ans. 39° 32′ 35″ N.

8. August 12th, in long 114° 35′ E, the obs mer alt $\underline{\odot}$ was 62° 21′ 10″, (ZS), IE + 4′ 20″; height of the eye 15 feet: find the latitude.

Ans. 12° 28′ 31″ S.

9. September 7th, in long 71° 20′ W, the obs double mer alt $\overline{\overline{\odot}}$ was 115° 42′ 30″, (ZN), IE — 2′ 20″: find the latitude.

Ans. 38° 12′ 58″ N.

ELEMENTS FOR PRECEDING EXAMPLES FROM NAUTICAL ALMANAC, 1876,

Declination.	Change in 1 hour.	Semi-diameter.
1. S 22° 0′ 57″·7	— 22″·08	16′ 18″·1
2. S 15 24 34 ·2	— 46 ·68	16 15 ·
3. S 3 27 10 ·9	— 58 ·92	16 7 ·5
4. N 9 38 12 ·1	+ 53 ·73	15 58 ·2
5. N 16 8 59 ·2	+ 43 ·02	15 53 ·2
6. N 22 49 21 ·9	+ 13 ·92	15 47 ·4
7. N 20 34 15 ·1	— 28 ·63	15 46 ·6
8. N 15 6 40 ·1	— 44 ·87	15 49 ·4
9. N 5 51 10 ·7	— 56 ·35	15 55 ·1

EXAMPLES FOR EXERCISE.

1. January 12th, 1876, the mer alt of Algenib was 51° 30′ 10″ (ZS), IE — 2′ 10″; height of the eye 14 feet: find the latitude.

Ans. 24° 6′ 46″ S.

2. February 8th, the mer alt Aldebaran was 57° 24′ 30″, (ZS), IE — 40″; height of the eye 17 feet; find the latitude.

Ans. 16° 25′ 13″ S.

3. March 10th, the mer alt Rigel was 62° 12′ 30″ (ZN) IE — 20″; height of the eye 14 feet; find the latitude.

Ans. 19° 31′ 13″ N.

4. April 14th, the mer alt Sirius was 72° 30′ 10″ (ZS, IE + 45″; height of the eye 16 feet; find the latitude.

Ans. 34° 6′ 14″ S.

5. May 7th, the mer alt Procyon was 62° 30′ 30″ (ZN), IE — 20″; height of the eye 17 feet; find the latitude.

Ans. 33° 6′ 52″ N.

6. June 7th, the mer alt Regulus was 39° 20′ 30″ (ZS), IE + 20″; height of the eye 15 feet; find the latitude.

Ans. 38° 9′ 49″ S.

NB.—* In questions 5 and 9, the altitude must be divided by 2, after applying Index Error There is no dip with an artificial horizon.

ELEMENTS FOR PRECEDING EXAMPLES FROM NAUTICAL ALMANAC, 1876.

Star.	Declination.
1. Algenib.	14° 29′ 41″ N
2. Aldebaran.	16 15 38 N
3. Rigel.	8 20 49 S
4. Sirius.	16 32 55 S
5. Procyon.	5 32 29 N
6. Regulus.	12 34 20 N

EXAMPLES FOR EXERCISE.

1. January 20th, 1876, in long 75° 20′ E, the obs mer alt $\underline{\odot}$ below pole was 5° 30′ 30″, IE + 50″; height of the eye 27 feet; find the latitude.
Ans. 77° 53′ 1″ S.

2. February 12th, in long 81° 30′ W, the obs mer alt $\underline{\odot}$ below pole was 3° 37′ 10″, IE − 2′ 10″; height of the eye 26 feet; find the latitude.
Ans. 79° 59′ 35″ S.

3. March 6th, in long 75° 10′ E, the obs mer alt $\overline{\odot}$ below pole was 2° 34′ 20″, IE + 10″; height of the eye 15 feet; find the latitude.
Ans. 86° 40′ 26″ S.

4. April 20th, in long 140° 30′ W, the obs mer alt $\overline{\odot}$ below pole was 5° 21′ 20″, IE — 20″; height of the eye 17 feet; find the latitude.
Ans. 82° 49′ 23″ N.

5. May 27th, in long 75° 30′ E the obs mer alt $\underline{\odot}$ below pole was 1° 37′ 15″, IE + 30″; height of the eye 14 feet; find the latitude.
Ans. 70° 1′ 58″ N.

6. June 14th, in long 144° 15′ W, the obs mer alt $\underline{\odot}$ below pole was 6° 21′ 20″, I.E — 1′ 40″; height of the eye 12 feet; find the latitude.
Ans. 73° 3′ 6″ N.

ELEMENTS FOR PRECEDING EXAMPLES FROM NAUTICAL ALMANAC, 1876.

Declination.	Change in 1 hour.	Semi-diameter.
1. S 17° 45′ 4″·1	— 40″·94	16′ 16″·2
2. S 13 48 5 ·9	— 49 ·72	16 14
3. S 5 24 24 ·1	— 58 ·23	16 8 ·8
4. N 11 44 7 ·1	+ 51 ·12	15 56 ·6
5. N 21 24 26 ·9	+ 24 ·55	15 48 ·7
6. N 23 18 27	+ 6 ·82	15 46 ·7

EXAMPLES FOR EXERCISE.

1. January 12th, 1876, the obs mer alt β Hydri below pole was 30° 20′ 30″, IE—2′ 10″; height of the eye 20 feet; find the latitude.
Ans. 42° 15′ 7″ S.

2. February 18th, the obs mer alt Achernar below pole was 10° 31′ 10″, IE + 45″; height of the eye 17 feet; find the latitude.
Ans. 42° 30′ 40″ S.

3. March 24th, the obs mer alt a Persei below pole was 13° 20′ 30″, IE — 27″; height of the eye 13 feet; find the latitude.
Ans. 53° 47′ 14″ N.

4. April 7th, the obs mer alt β Tauri below pole was 5° 37′ 20″, IE + 40″; height of the eye 15 feet; find the latitude.
Ans. 66° 54′ 57″ N.

ELEMENTS FOR PRECEDING EXAMPLES FROM NAUTICAL ALMANAC, 1876.

Star.	Declination.
1. β Hydri.	77° 57′ 12″ S
2. Achernar.	57 52 11 S
3. a Persei.	49 25 20 N
4. β Tauri.	28 30 12 N

5. TO FIND THE LATITUDE BY A MERIDIAN ALTITUDE OF THE MOON.

The moon being the nearest heavenly body to the earth, her distance from it being only 60 times the earth's radius, and her diameter 2153 miles, her change of declination is more rapid than that of any other heavenly body; her parallax too is large, and her change of semi-diameter in a given time considerable. These things necessitate careful corrections of all her elements, and render observations of that planet more difficult to deal with than those of others. The method of finding the latitude by a meridian altitude of the moon, is a somewhat complicated problem, and to the beginner of astronomy presents considerable difficulty on account of the exercise of memory required. It might therefore be properly placed as a more advanced problem, but as it is a meridian altitude we are about to work upon, its place must be under this heading.

This method is not much used at sea, on account of the detailed corrections to be applied, and is hardly likely to become a popular method when easier processes are within reach of the navigator. Still, it is useful to verify results obtained in other ways when opportunity offers.

Note.—(a). Find the Greenwich date of the moon's meridian passage. (See Subordinate Computation, E.)

(*b*) Take out her semi-diameter, and horizontal parallax for that Greenwich date. See Subordinate Computation, D.)

(*c*) Correct the moon's declination for that date. (See Subordinate Computation, F.)

(*d*) Correct the altitude as far as refraction.

(*e*) Fill in the augmentation for the moon's semi-diameter from table; always added, entering the table with the altitude roughly and the semi-diameter.

(*f*) Find the latitude nearly by subtracting the altitude from 90° and applying the declination.

(*g*) From table of reduction of moon's parallax take out the reduction, (always subtracted from the parallax) entering the table with latitude nearly and horizontal parallax. Reduce the horizontal parallax to seconds.

(*h*) Compute the parallax in altitude with the formula, parallax in altitude "= HP" cos app alt., which add to the altitude, and find the latitude as before.

EXAMPLE.

January 18th, 1876, in longitude 32° 10′ W, the obs mer alt ☾ was 64° 30′ 30″, (ZS), IE + 2′ 10″; height of the eye 17 feet: find the latitude.

Greenwich Date.	Retardation.	Longitude.
d h m	m	° ′
Mer pass 17 17 44·7	+ 43·1	32 10 W
Retn. + 3·7	2·1	4
17 17 48·4	431	6,0)12,8 40
Long + 2 8·7	862	2 8 40
17 19 57·1	4)90·51	
	6)22·6	
	3·7	

☾ Semi-diam.	Change.	☾ Hor Parx.	Change.
′ ″	″	′ ″	″
15 9·2	— 5·4	55 30·9	— 19·7
Change — 3·5	7·9	Change — 12·9	7·9
15 5·7	486	55 18·0	1773
Augmentn. + 13·5	378	Redn. — 3·8	1379
15 19·2	12)42·66	55 14·2	12)155·63
	— 3·4	60	12·9
		H.P. 3314″	

☾ Declination.	Change.
° ′ ″	″
S 12 20 49·3	+ 13·45
12 48·0	57· 1
12 33 37·3	1345
	9415
	6725
	6,0)76,7,995
	12·48

Parallax in Altitude.

Par^x. in alt = Hor Parallax cos app alt.
Hor^l. Par^x. 3314″ = 3·520353
Cos app alt 64° 43′ = 9·630524
6,0)141,5″ = 3·150877
Par^x. in alt + 23′ 35″

	°	′	″	
Obs alt $\overline{☾}$	64	30	30	
$\overline{\text{IE}}$ +		2	10	
	64	32	40	
Dip —		4	4	
	64	28	36	
Ref —			28	
	64	28	8	
SD +		15	19	
	64	43	27	
Parallax +		23	35	
True alt	65	7	2	
	90	0	0	
Zen dist	24	52	58	S
Declination	12	33	37	S
Latitude	37	26	35	S

EXAMPLES FOR EXERCISE.

1. February 18th, 1876, in long 67° 20′ E, the obs mer alt $\overline{☾}$ was 72° 30′ 30″ (ZN), IE — 40″; height of the eye 18 feet; required the latitude.

Ans. 9° 56′ 30″ S.

2. March 28th, in long 69° 20′ W, the obs mer alt $\underline{☾}$ was 75° 31′ 30″ (ZS), IE + 42″; height of the eye 19 feet; required the latitude.

Ans. 3° 38′ 40″ N.

3. April 12th, in long 76° 10′ E, the obs mer alt ☾ was 64° 30′ 30″ (ZN), IE — 25″ ; height of the eye 13 feet ; required the latitude.

Ans. 0° 11′ 3″ S.

4. May 16th, in long 29° 30′ E, the obs mer alt ☾ was 51° 20′ 20″ (ZS), IE + 30″; height of the eye 14 feet ; required the latitude.

Ans. 52° 50′ 15″ S.

5. June 13th, in long 81° 20′ W, the obs mer alt ☾ was 62° 40′ 30″ (ZN), IE —27″; height of the eye 16 feet; required the latitude.

Ans. 17° 27′ 28″ N.

6. July 15th, in long 27° 20′ W, the obs mer alt ☾ was 54° 21′ 10″ (ZS), IE + 40″; height of the eye 15 feet; required the latitude.

Ans. 16° 32′ 58″ S.

Elements for preceding Examples from Nautical Almanac, 1876.

	Meridian passage at Greenwich.	Retardation.	Declination.	Change in 10m
	d h m	m		
1.	18 7 8·2	49·9	S 27° 27′ 20″·9	+ 44″·11
2.	27 13 49·3	47·0	N 17 31 36 ·1	+139 ·29
3.	12 2 50·6	49·2	S 25 32 3 ·8	+ 70 ·65
4.	16 6 27·0	44·3	S 15 1 41 ·5	—128 ·11
5.	13 5 6·0	42·0	S 9 25 52 ·3	—140 ·65
6.	15 6 44·1	46·2	N 18 11 27 ·6	+131 ·21
	Semi-diameter.	Change.	Horizontal Parallax.	Change.
1.	14′ 47″·5	—·″6	54′ 11″·6	— 2″·1
2.	16 7 ·5	+2·4	59 4 ·5	+ 8 ·8
3.	14 56 ·9	—2·9	54 46 ·1	—10 ·7
4.	15 5 ·2	+5·4	55 16 ·6	+19 ·6
5.	15 4 ·5	+5·1	55 14 ·0	+18 ·6
6.	15 54 ·1	+7·1	58 15 ·6	+26 ·1

(6) To find the Latitude from an Altitude of the Pole Star.

The elevation of the Pole above the plane of the observer's horizon measures the latitude of the place: (see Investigation of method by Meridian Altitude): hence, if a star or other heavenly body were fixed at the pole, its altitude, corrected for Index Error, Dip, &c., would be the latitude. The nearest heavenly body to the north pole is the star *α* Polaris, commonly known as the Pole Star: this object is, of course, circum-polar, *i.e.*, its polar distance is less than the elevation of the pole above the plane of the horizon, which is the latitude: hence it never sets.

Fig 18.

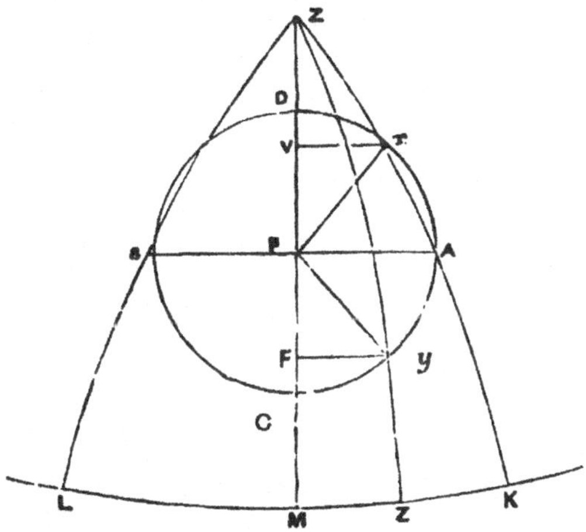

Let LK represent the observer's horizon: ZM his meridian, Z the zenith, P the pole, and ACB the daily circle described by the pole star.

It is evident that when the star is in either of the positions A or B its altitude AK or BL will nearly equal the elevation of the pole PM above the horizon LMK: *i.e.*, when its hour angle DPA is 6^h or a right angle, or again when the star has travelled to B, and its hour angle is 18^h. In that case all corrections to be applied to the altitude but its own, (*e.g.*, index error), will vanish. Again in the position D, the star will be at its upper transit, and PM (lat) = DM (star's alt) — PD (star's polar dist). In the position C, the body will be at its lower transit, or on the meridian below pole, when PM (lat) = PC (star's polar dist) + CM (star's alt).

When the star is in any other position, *e.g.*, as at *x* when its hour angle lies between 0^h and 6^h, then *x*K (its altitude) — PV (the correction) = PM (the latitude:) and in the position *y*, *y*Z (its altitude) + PF (the correction) = PM (the latitude).

Hence the following practical rule, when the star is anywhere on the semi-circle ACB, the correction must be added to the altitude, and in the upper half-circle BDA, subtracted, or if the hour angle lies between 6^h and 18^h, the principal correction is + : between 0^h and 6^h —; and between 18^h and 24^h—.

We must defer the mathematical investigation of the method, introducing here only the above explanation of the principle upon which it depends, and now proceed to its practical working.

Example.—On January 5th, 1876, at 2^h 30^m 30^s A.M., mean time, in longitude 37° 10′ W, the obs alt of Polaris, was 51° 44′ 20″, IE — 5′ 0″ : height of the eye 15 feet : find the latitude.

Note.—(*a*) Find the Greenwich Date from the above time.

(*b*) Correct the Mean Sun's Right Ascension : take out the Star's Right Ascension and Declination for the Greenwich Date : find the polar distance and reduce it to seconds : (call this *p*).

(*c*) Find the hour angle (subordinate computation 1).

(*d*) Find the first and second corrections by the formulæ (p cos h), and ½ sin 1″ (p sin h)² tan alt which applied to the true altitude gives the latitude.

Greenwich Date.

d	h	m	s
4	14	30	30
	2	28	40
4	16	59	10

Longitude.

	°	′
	37	10 W
		4
6,0)	14,8	40
	2	28·40

Mean Sun's right Asⁿ

h	m	s
18	53	47·78
	2	37·70
		9·69
		·03
18	56	35·20

Star's right Asⁿ.

h	m	s
1	13	2·25

Star's polar dist.

	°	′	″
	88	39	16
	90	0	0
	1	20	44
	60		
	80		
	60		
(*p*)	4844		

Hour angle.

	h	m	s
MTP	14	30	30
Mean Sun's RA +	18	56	35·2
	33	27	5·2
Star's RA —	1	13	2·2
	32	14	3·0
Reject	24	0	0
Hour angle	8	14	3
	60		
4)	494	3	
(H)	123	30·45	

1st Correction.

(*p* cos *h*)

p 4844″	log	3·685204
cos H	log	9·741889
6,0)267,3″	=	3·427093
44, 33″		

2nd Correction.

(*p* sin *h*)² tan alt ½ sin 1″

p 4844″	log	3·685204
sin H	log	9·921107
		3 606311
		2
		7·212622
Tan alt		10·100432
½ sin 1″		4·384545
49″·84 =		1·697599

Latitude.

	°	′	″
Obs alt	51	44	20
IE —		5	0
	51	39	20
Dip —		3	49
	51	35	31
Ref —			45
True alt	51	34	46
1st cor +		44	33
	52	19	19
2nd cor +			49
Latitude	52	20	8

N.B.—The first correction is added because the hour angle 8ʰ is between 6ʰ and 18ʰ. (See foregoing explanation.) The second correction is always to be added.

The same Example worked by a shorter method.

The pole star tables at the end of the Nautical Almanac facilitate the work, and give a result, which, though approximate in some instances, is useful at sea to verify the latitude by another method. These tables are to be found in the Almanac for 1876, at pages 493—495. One minute should always be subtracted from the altitude in this method.

Mean Time nearly		14^h 30^m
Sid Time	+	18 56
		33 26
	reject	24 0
Right as^n^ of mer^n^		9 26

	True alt	51° 34′ 46″		52° 20′ 53″
TABLE I.	with the sidereal time, *i.e.* RA of mer	+ 44 25	Subtract	1 0
			Latitude	52 19 53
TABLE II.	with the alt and RA of mer	+ 47		
TABLE III.	with day and RA of mer	+ 55		
		52 20 53		

If the tables be carefully entered, *i.e.*, the intermediate minutes worked between those given, the result will be sufficiently exact for practical purposes.

EXAMPLES FOR EXERCISE.

1. February 9th, 1876, at 1^h 49^m 40^s a.m., mean time, in long 42° 15′ W the obs alt of the Pole Star was 57° 32′ 10″, IE + 2′ 10″; height of the eye 14 feet: find the latitude.

Ans. 58° 38′ 52″ N.

2. March 12th, at 10^h 20^m 40^s P.M., mean time, in long 32° 10′ E, the obs alt of the Pole Star was 64° 12′ 30″, IE — 40″; height of the eye 12 feet: find the latitude.

Ans. 64° 58′ 50″.

3. April 18th, at $9^h\ 37^m\ 30^s$ P.M., mean time, in long 39° 44′ E, the obs alt of the Pole Star was 59° 29′ 30″, IE — 30″; height of the eye 17 feet: find the latitude.

Ans. 60° 37′ 16″.

4. May 7th, at $6^h\ 42^m\ 10^s$ P.M., mean time, in long 116° W, the obs alt of the Pole Star was 60° 44′ 10″, IE + 5″; height of the eye 18 feet: find the latitude.

Ans. 61° 31′ 19″.

ELEMENTS FOR PRECEDING EXAMPLES FROM NAUTICAL ALMANAC, 1876.

	Sidereal Time.	Star's RA.	Star's Dec.
	h m s	h m s	° ′ ″
1.	21 11 47·29	1 12 30·74	88 39 15
2.	23 21 53·6	1 12 10·58	88 39 8
3.	1 47 46·1	1 12 8·0	88 38 57
4.	3 2 40·6	1 12 15·62	88 38 52

(7). LATITUDE BY REDUCTION TO THE MERIDIAN.

This method consists in reducing an altitude of a heavenly body, when near the meridian, to that which it would have when on the meridian. The reduction, therefore, is the difference between the two altitudes. When a heavenly body is in the same hemisphere with the observer, *i.e.*, when its declination is of the same name as his latitude, then its meridian altitude is its greatest altitude: but if in the opposite hemisphere, the greatest altitude may be reached either before or after the object is on the meridian of the place. The difference in the altitudes, however, is so small as to be inappreciable in the case either of the sun or a star. But with the moon, owing both to her proximity to the earth, and to her very unequal motion in the heavens, caused by the combined influence of the earth's and sun's attraction, when she lies in a contrary hemisphere her meridian altitude differs considerably from her greatest altitude. Hence, observations of the moon as ex-meridian altitudes in such a case, are useless. In the case of the sun, when the sky is obscured as he makes his transit, and the meridian altitude cannot be obtained, this "Reduction" method is very useful. An error in a single observation, such as a meridian altitude, will vitiate the result: but as the sun approaches the meridian, a series of altitudes may be deliberately and carefully taken, and thus the chances of error are diminished. These altitudes will change slowly. The error of the chronometer on apparent time at place must be accurately known, having been previously obtained from an earlier set of observations. It may be sufficient here to state that, the nearer the meridian, the better the observation for latitude, while the reverse holds good in observations for finding the time. (See postea.)

Fig. 19.

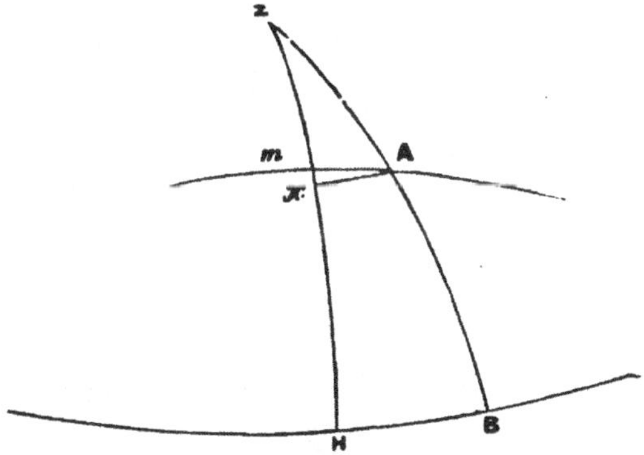

In the accompanying figure, let A represent the place of the body when near the meridian at the time of observation, and *m* its place when on the meridian. It is evident, that the altitude *m*H is greater than AB by *mx*, which is the Reduction, and must be added to AB to obtain *m*H. Hence the reduction is either added to the true altitude to obtain the meridian altitude; or it is the difference between the Zenith distance at the time of observation (ZA), and the meridian zenith distance (Z*m*), in which case, it is subtracted from the former.

EXAMPLE I.

When the Error of Chronometer is known for Apparent Time at Place.

Note.—(*a*) Find the means of the times and altitudes, and applying the error and longitude to the former gives the true apparent Greenwich date.

(*b*) With this date take out the sun's declination.

(*c*) Find the chronometer at noon, by applying the error to 12^h; *i.e.*, find what the chronometer actually showed, when it was apparent noon at the place.

(*d*) The difference between this and each of the times gives the hour angles, the values for which fill in from table; the mean of these M.

(*e*) The meridian zenith distance, found by applying the latitude to the declination, if of contrary names add.

(*f*) Compute the reduction : $r'' =$ cos dec cos lat cosec MZD log M always + to the true altitude.

(*g*) Correct the altitude and apply reduction : find the zen dist, and applying dec gives the latitude.

January 10th, 1876, in latitude by account 39° 15′ S, longitude 32° E, the following observations were taken near noon to determine the latitude.

Chronometer.	Sextant.
h m s	° ′ ″
9 31 10	☉ 72 30 30
32 10	31 10
32 59	31 50
33 40	32 20
34 10	32 50

Index Error — 20″; height of the eye 17 feet: chronometer slow on apparent time at place 2^h 30^m 30^s.

	h m s		Longitude.	° ′ ″
	9 31 10		32 E	72 30 30
	32 10		4	31 10
	32 59		6,0)12,8	31 50
	33 40		2^h 8^m	32 20
	34 10			32 50
	5)164 9			5)158 40
	9 21 32 49·8		Mean of alts	72 31 44
Error Chron	+ 2 30 30	slow		
True App Time	10 0 3 19·8			
Longitude	— 2 8 0			
Greenwich Date	9 21 55 19·8			

Declination.	Change.	Chron at Noon.	
° ′ ″	″		h m s
S 22 9 35	—21·00	App noon	12 0 0
— 7 40	21· 9	Chron slow	2 30 30
22 1 55	18900	Chron shows	9 29 30
	2100		
	4200		
	6,0)45,9·900		
	7 39·9		

Hour Angles.			Reduction.	
m s			r'' = cos dec cos lat cosec MZD log M	
1 40	=	5	Cos dec	9·967064
2 40	=	14	Cos lat	9·888961
3 29	=	24	Cosec MZD	10·528321
4 10	=	34	Log M	1·380211
4 40	=	43	R = 58″.1 =	1·764557
		5)120		
M	=	24		

Mer Zen Dist.

	°	′	
Lat	39	45	S
Dec	22	1	S
Diff =	17	44	S

Reduction.

	°	′	″	
Mean of alt ☉	72	31	44	
IE	—		20	
	72	31	24	
Dip	—	4	4	
	72	27	20	
Ref	—		18	
	72	27	2	
SD	+	16	18	
	72	43	20	
Par^x	+		3	
True alt ☉	72	43	23	
Red^n.	+		58	
Mer alt	72	44	21	
	90			
Zen dist	17	15	39	S
Dec	22	1	55	S
Latitude	39	17	34	

EXAMPLE II.

When the Error of Chronometer is known for Mean Time at Place.

In the preceding Example, suppose the chronometer to be slow on Mean Time at place $2^h\ 30^m\ 30^s$, then the Greenwich date will be mean time, the declination the same, also the meridian zenith distance. The equation of time is now required.

Equation Time.

	m	s
	7	13·6
		22·7
	7	36·3

+ to app time

Chronometer.

s
+1·039
21·9
9351
1039
2078
22·7541

Chronometer at Noon.

	h	m	s
	12	0	0
Eq T +		7	36
Mean T	12	7	36
Error on M T	2	30	30
	9	37	6

Hour Angles.

m	s		
5	36	=	62
4	56	=	48
4	7	=	33
3	26	=	23
2	56	=	17
			5)183
		M =	36·6

Reduction.		Latitude.				
			°	′	″	
Cos dec	9·967064	True alt ☉	72	43	23	
Cos lat	9·888961	Redⁿ·	+	1	29	
Cosec MZD	10·528321	Mer alt	72	44	52	
Log M	1·563481		90			
88″·7 =	1·947827	Zen dist	17	15	8	S
		Dec	22	1	55	S
		Latitude	39	17	3	S

EXAMPLE III.

When the Error of Chronometer is known for Greenwich Mean Time.

January 12th, 1876, in lat by acc 30° S, longitude 40° E, the following observations were made.

Error of Chronometer on Greenwich mean time 0^h 56^m slow.

Times.		Altitudes.
h m s		° ′ ″
8 30 20		☉ 81 30 0
31 10		31 10
32 30		32 30

Index Error—2′ 10″. Height of the eye 27 feet: find the latitude.

	d	h	m	s		°	′	″
Mean of times	11	20	31	20	Mean of alts	81	31	13
Error slow		+	56	0				
Greenwich mean time	11	21	27	20				

Declination.	Change.	Equation of Time.	Change.
° ′ ″	″	m s	
S 21 51 58	— 23·14	8 2·36	+ ·992
— 8 17	21· 5	+ 21·33	21·5
21 43 41	11570	8 23·69	4960
	2314	+ AT	992
	4628		1984
	6,0)497,510		21·3280
	8′ 17″·5		

Chron at noon.

	h	m	s
	12	0	0
Equation time +		8	23·7
	12	8	23·7
Error −		56	0·0
	11	12	23·7
Long −	2	40	0
Chron sh[d] at noon	8	32	23·7

Longitude.

40° E
4
6,0)16,0
$2^h\ 40^m$

Hour Angles.

m	s	
2	3	= 8
1	13	= 3
0	7	= 0
		3)11
		M = 3·66

Mer zen dist.

Lat	30° 0′	S
Dec	21 44	S
MZD	8 16	S

Reduction.

Cos dec	9·967977
Cos lat	9·937531
Cosec MZD	10·842300
Log M	0·563481
20″·5 =	1·311289

Latitude.

	°	′	″	
Mean of alts ☉	81	31	13	
IE −		2	10	
	81	29	3	
Dip −		5	7	
	81	23	56	
Ref −			8	
	81	23	48	
SD +		16	18	
	81	40	6	
Parallax +			2	
True alt	81	40	8	
Red[n] +			20	
Mer alt	81	40	28	
	90	0	0	
Zen dist	8	19	32	S
Dec	21	43	41	S
Latitude	30	3	13	S

EXAMPLE IV.

When the hour angles exceed 20^m (for which the table is computed), the reduction must be worked as follows, which renders the calculation, in any case, independent of the tables; *e.g.*, let

Mean of Times be January	12^d	20^h	30^m	30^s
And Error on app time slow	+	3	0	0
True app time at place	12	23	30	30

To find Hour Angle.

True app time	23^h	30^m	30^s
Subtract from	24	0	0
	0	29	30
	60		
	4)29	30	
	2)7	22	30
$\frac{1}{2}h =$	3°	41′	15″

Value of M $= \dfrac{2 \sin^2 \frac{1}{2}h}{\text{Sin } 1''}$

$\frac{1}{2}h$ 3° 41′ 15″ sin	8·808308
	2
	17·616616
Log 2 +	0·301030
	17·917646
(Constant) Sin 1″ —	4·685575
$\therefore \dfrac{2 \text{ Sin}^2 \frac{h}{2}}{\text{Sin } 1''}$	13·232071

Reduction.

Suppose lat 39° cos	9·890503
Suppose dec 22° cos	9·967166
Suppose MZD 61 cosec	10·058181
$\dfrac{2 \text{ Sin}^2 \frac{1}{2}h}{\text{Sin } 1''}$ . .	13·232071
6,0)140,6″ =	3·147921
+ 23′ 26″	

∴ Reduction = 23′ 26″.

The remainder of the work proceeds as before.

EXAMPLES FOR EXERCISE.

1. February 18th, 1876, in lat by account 42° 30′ N, longitude 17° W, the following observations were taken near noon to find the latitude.

Chronometer.	Sextant.
h m s	° ′ ″
2 41 27	$\underline{\odot}$ 35 48 10
42 30	49 10
43 44	49 50
44 50	50 20
46 10	50 50

Index Error + 45″; height of the eye 13 feet; chronometer fast on app time at place 2^h 30^m 30^s.

Ans. 42° 9′ 17″ N.

2. March 17th, 1876, in lat by account 34° 30′ N, longitude 39° E.

Chronometer.	Sextant.
8h 31m 27s	$\underline{\odot}$ 54° 40′ 10″
32 40	39 20
33 50	38 50
34 40	38 0
35 30	37 20

Index Error — 22″: Height of the eye 16 feet. Error of Chronometer on Mean Time at place 3h 30m 14s slow,

Ans. 34° 33′ 16″ N.

3. April 17th, 1876, in lat by account 44° 30′ S, longitude 42° W.

Chronometer.	Sextant.
1h 27m 10s	$\underline{\odot}$ 34° 2′ 10″
27 50	2 50
28 30	3 30
29 0	4 0
29 50	4 50

Index Error + 1′ 30″; height of the eye 14 feet. Error of Chronometer on Greenwich Mean Time 1h 21m 30s slow.

Ans. 45° 31′ 21″ S.

ELEMENTS FOR PRECEDING EXAMPLES FROM NAUTICAL ALMANAC, 1876.

1. Dec noon, February 18th, S 11° 44′ 54″; change in 1h — 52″·83; semi-diameter 16′ 13″.

2. Dec noon, March 16th, S 1° 28′ 57″; change in 1h — 59″·25; equation of time 8m 38s·4; change in 1h — ·729s to be added to apparent time; semi-diameter 16′ 6″.

3. Dec noon, April 17th, N 10° 41′ 57″; change in 1h + 52″·48: equation of time 0m 34s·92; change in 1h + ·579s to be subtracted from apparent time; semi-diameter 15′ 57″.

(8) LATITUDE BY ANY EX-MERIDIAN ALTITUDE.

The Latitude may be found from the observation of a heavenly body at any time, provided only that the error of the Chronometer is exactly known; *i.e.*, that the exact time, (mean or apparent) at the place of observation when the altitude or altitudes are taken, is correctly found. This method is seldom, if ever, used at sea through the absence of the above, or possibly, because it is not generally known. It may, however, be regarded as working the latitude backwards from the error of chronometer, the reverse of that described under the heading "Longitude by Chronometer," in which, when the latitude is accurately known, the true mean or apparent time is found. It may be worth while, however, to describe it as a possible method under the heading of "Latitude."

EXAMPLE.

January 10th, 1876, in North latitude, longitude 72° 20′ W, at $1^h\ 20^m\ 10^s$ P.M., mean time at place, the observed alt ☉ was 22° 30′ 10″, IE + 1′ 30″; HE 24 feet: find the latitude.

Greenwich Date.

d	h	m	s
10	1	20	10
	4	49	20
10	6	9	30

Longitude.

	°	′	
	72	20	W
		4	
6,0)	28,9	20	
	4	49	20

Declination.

	°	′	″
	S 22	1	0·5
	—	2	14·7
	21	58	45·8
	90	0	0
Polar dist	111	58	46

Change.

″
—22·08
6· 1
2208
13248
6,0)134,688
—2 14 7

Equation of Time.

m	s
7	38·26
	6·20
7	44·46
—Mean Time.	

Change.

s
+1·016
6·1
1016
6096
6·1976

Hour Angle.

	h	m	s
Mean Time	1	20	10
Equation Time	—	7	44·5
	1	12	25·5
	60		
	4)72	25	30
Hour Angle = H	18	6	22

Corrected Altitude.

	°	′	″
Obs alt ☉	22	30	10
IE	+	1	30
	22	31	40
Dip	—	4	49
	22	26	51
Ref		2	17
	22	24	34
SD	+	16	18
	22	40	52
Parallax	+		8
True alt ☉	22	41	0
	90	0	0
True zen dist	67	19	0

COMPUTATION.

In the accompanying Fig. call the Zenith Distance ZS, the Polar distance PS, and the Hour Angle ZPS, *h*. From S, which is the position of the sun at the time of observation, draw SQ perpendicular to PZ produced.

Fig. 20.

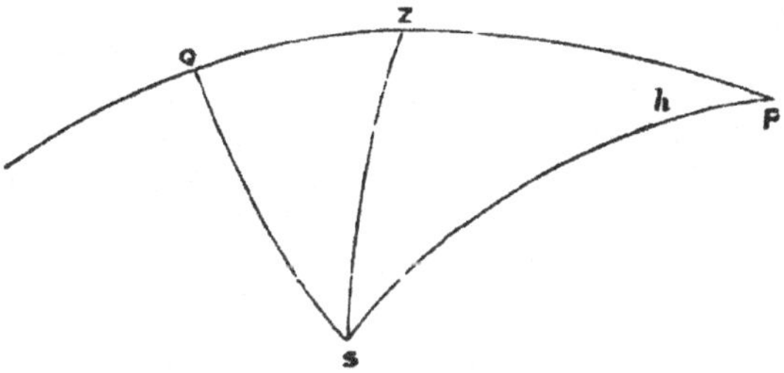

Then by Napier's rules for the solution of right angled spherical triangles the following formulæ are obtained—

(1) Sin SQ = Sin PS Sin h (2) Cos PS = Cos PQ Cos QS (3) Cos ZS = Cos ZQ Cos QS

∴ Cos PS Sec QS = Cos PQ ∴ Cos ZS Sec QS = Cos ZQ.

Sin PS 9·967222	Cos PS 9·573189	Cos ZS 9·586179
Sin h 9·492449	Sec QS 10·018827	Sec QS 10·018827
Sin 9·459671	Cos 9·592016	Cos 9·605006
SQ = 16° 44′ 58″	PQ = 66° 59′ 32″	ZQ = 66° 15′ 6″
	180 0 0	
	113 0 28	

(4) PQ − ZQ = PZ (the colatitude)

PQ =	113° 0′ 28″
ZQ =	66 15 6
Co lat =	46 45 22
	90 0 0
Lat =	43 14 38 N

N.B.—PQ is greater than 90°, when PS, the polar distance is more than 90°.

This method answers well for the latitude, when the hour angles are beyond the limits of Reduction Tables.

(9) Ivory's Method of Double Altitudes.

This method is founded upon the supposition that two altitudes of a heavenly body are taken and the interval between the observations is carefully noted, but more especially, no allowance is made for change of polar distance in that interval. It answers well for two observations of the same star, when the polar distance may be said not to change; in the case of the sun, if we use the mean polar distance, or that at the middle time between the observations; and also for the moon when she is in extreme N or S declination, and when, therefore, her change of polar distance is least rapid. Ivory's method may be said chiefly to consist in the solution of a number of small right angled spherical triangles, from which the co-latitude is found.

The best species of observation to which to apply Ivory's method, is that of two altitudes of the sun, taken, one in the morning, the other in the afternoon with such an interval, that the middle time shall be as near as possible to noon. This method is invaluable at sea, and is of easy application to verify other results, and therefore should be often used.

EXAMPLE.

February 27th 1876, in South lat, long 39° 40′ W, the following observations were taken to determine the latitude.

Mean Times nearly	Chron Times	Obs alts ☉
h m	h m s	° ′ ″
8 37 A.M	9 57 9	32 42 10
2 39 P.M	4 8 56	43 1 48

Index Error —1′ 13″. Height of the eye 19 feet. Sun's bearing at the first observation SE b E½E ; ship's run in the interval SSW¼W, 5 miles an hour.

Note.—(*a*) Find the Middle Greenwich Date from the mean times at place by applying the longitude to their mean.

(*b*) Find half the elapsed chronometer time, and reduce this to arc.

(*c*) Find the Sun's polar distance for the Middle Greenwich Date.

(*d*) The correction for ship's run, found by multiplying the dist run in the interval, by the interval between the observations ; adding to the log of this number the cosine of the angle between the sun's bearing and ship's run. The result is in minutes of arc, and is to be added to the first altitude when the angle is under 8 points, otherwise subtracted.

(*e*) Correct the altitudes, and find their half sum and difference (after applying the correction for ship's run) : call the former S, and the latter D.

(*f*) Work the latitude by Ivory's method.

	Middle Greenwich Date.	Longitude.		Half Elapsed Time.
	d h m s	° ′		h m s
Mean Times	26 20 37 0	39 40 W	Chron	9 57 9
,,	27 2 49 0	4	,,	16 8 56
	2)53 23 26 0	6,0)15,8 40		2)6 11 47
	26 23 43 0	2 38 40		3 5 53·5
Longitude	+ 2 38 40			60
Greenwich Date	27 2 21 40			4)185 53 30
			H	46° 28′ 22″

Sun's Polar Dist.

	°	′	″
S	8	28	12
		2	9
	8	26	3
	90	0	0
p	81	33	57

Change.

″
−56·32
2·3
16896
11264
6,0)129,536
2 9·5

Ship's Run.

Dist run = 5 × 6^{h}·2 = 31′ viz: 5 miles per hour × the diff of chron times.

And in the following fig.:

Fig. 21.

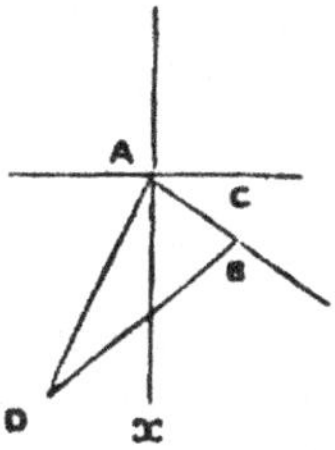

*x*AB = 5½ points being sun's bearing SE by E ½ E; *x*AD = 2¼ points for ship's run is SSW ¼ W ∴ DAB = 2¼ + 5½ = 7¾ points.

Then

AB (Cor) = AD dist run × cos 7¾ points.
= 31 × cos 7¾ points.

Log 31 = 1·491362
Log cos 7¾ = 8·690796
Cor + 1′·52   0·182158

Corrected Altitudes.

	32°	42′	10″
IE	−	1	13
	32	40	57
Dip	−	4	17
	32	36	40
Ref	−	1	29
	32	35	11
SD	+	16	11
	32	51	22
Parallax	+		7
	32	51	29
Cor Ship's run	+	1	30
1st alt	32	52	59
2nd alt	43	11	34
	2)76	4	33
S	38	2	16
	2)10	18	35
D	5	9	17

	43°	1′	48″
IE	−	1	13
	43	0	35
Dip	−	4	17
	42	56	18
Ref	−	1	2
	42	55	16
SD	+	16	11
	43	11	27
Parallax	+		7
	43	11	34

To find the Latitude.

H	46° 28′ 22″	sin 9·860366				
p	81 33 57	sin 9·995278	cos	9·166351		
(A)	45 49 27	sin 9·855644	sec	10·156853	cosec	10·144357
		E 77° 50′ 59″	cos	9·323204		

		Sec A 10·156853		
		9·789708	sin S cos	9·896309
		9·998240	cos D sin	8·953496
		10·002124	sec sin	8·994162
(O)	27° 45′ 10″	cos 9·946925	(I)	5° 39′ 44″
(E—O)	50 5 49	cos 9·807191		
(I)	5 39 44	cos 9·997876		
Lat	39 40 11	sin 9·805067		

N.B.—(E) is always of the same name as *p*, *i.e.*, when *p* is more than 90° (E) must be taken from 180°. It may happen within the tropics that the fifth arc is (E + O), *i.e.*, when the lat is less than 23° 28′.

EXAMPLES FOR EXERCISE.

1. March 8th, 1876, in North lat, long 27° 20′ W, the following observations were taken to find the latitude by Ivory's method.

Mean Times nearly.	Chron Times.	Obs Alts ☉
8ʰ 27ᵐ A.M.	9ʰ 51ᵐ 45ˢ	14° 38′ 45″
2 24 P.M.	3 48 45	24 35 37

Index Error — 1′ 19″: height of the eye 19 feet. Sun's bearing, first observation, SE ¾ E: ship's run in the interval, S b W ¼ W, 3·5 miles an hour.

Ans. 54° 33′ N.

2. April 20th, in North latitude, longitude 17° 10′ W,

Mean Times nearly.	Chron Times.	Obs Alts ☉
2ʰ 44ᵐ P.M.	4ʰ 12ᵐ 30ˢ	40° 52′ 40″
5 21 P.M.	6 48 40	14 43 38

Index Error + 1′ 24″: height of the eye 20 feet. Sun's bearing, first observation, SSW ½ W: ship's run in the interval WSW ¾ W, 5 miles an hour.

Ans. 45° 26′ N.

3. August 28th, in South latitude ; long 84° 30′ E,

Mean Times nearly.	Chron Times.	Obs Alts $\overline{\odot}$
9ʰ 24ᵐ A.M.	6ʰ 6ᵐ 5ˢ	27° 22′ 30″
3 22 P.M.	12 4 15	20 57 40

Index Error + 1′ 10″: height of the eye 18 feet. Sun's bearing at first observation E: ship's run in the interval SSW, 5 miles an hour.

N.B.—When the angle between the sun's bearing and ship's run exceeds 8 points, use the supplement, *i.e.*, subtract from 16 points, and the correction will be subtractive.

Ans. 42° 28′ S.

ELEMENTS FOR PRECEDING EXAMPLES FROM NAUTICAL ALMANAC, 1876.

1. Dec noon March 8th, S 4° 37′ 51″; change in 1ʰ, − 58″·55 ; semi-diameter 16′ 8″.

2. Dec noon April 20th, N 11° 44′ 8″; change in 1ʰ, + 51″·12 ; semi-diameter 15′ 57″.

3. Dec noon Aug 27th, N 9° 52′ 1″, change in 1ʰ, − 52″·86 ; semi-diameter 15′ 52″.

For the longitude by Ivory's method, see longitude by chronometer : the corrected latitude must be used.

(10) DOUBLE ALTITUDES BY DIRECT SPHERICS.

This method is solved by the direct application of the rules of Spherical Trigonometry, the following elements being known, viz., the two polar distances of the same object, or the polar distance of each of two objects ; the two altitudes, and, therefore, the Zenith distances ; and the Polar angle between the two positions of the object.

Fig. 22.

Fig. 1. Fig. 2.

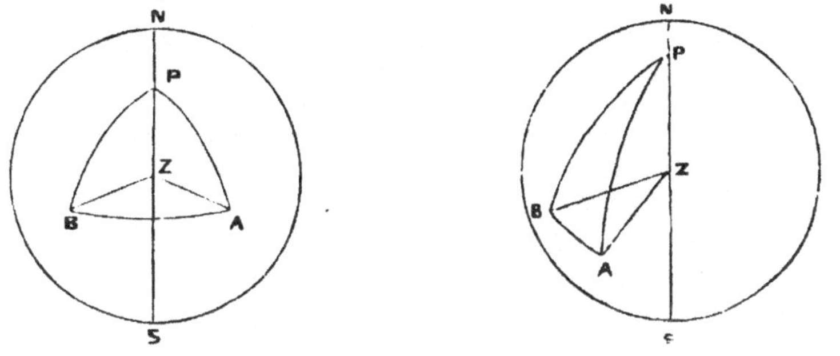

Fig. 3.

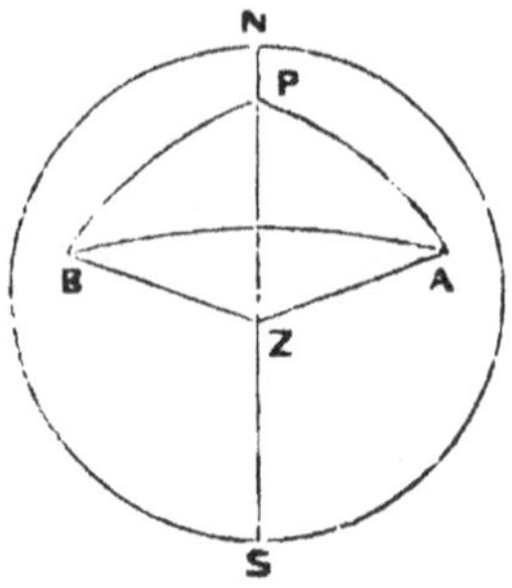

Fig. 4.

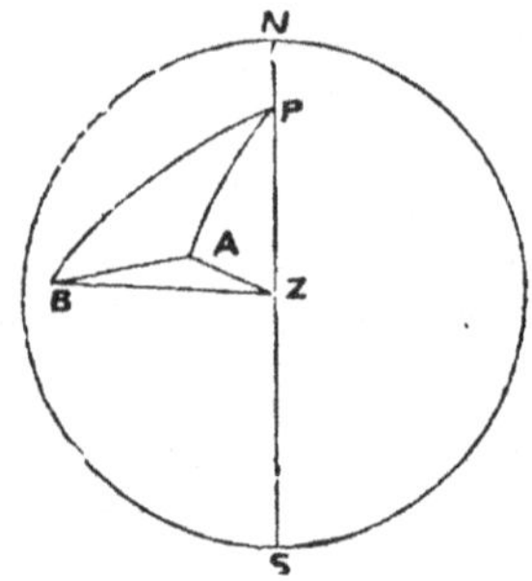

There are four possible cases of this problem which the accompanying figures will illustrate, depending upon the relative lengths of the polar distances and the colatitude of the place.

In figs. 1 and 3, the objects are represented as being on either side of the meridian NS. In fig. 1, the polar distances PA, PB are both greater than the colatitude PZ; in fig. 3 they are both less; therefore, in the former case, the great circle AB joining the positions of the objects A and B will fall below, and in the latter case, above the Zenith Z. In figs. 2 and 4. the bodies are represented as being both on the same side of the meridian: in the former, the nearest polar distance PA to the meridian NS, is greater than the more remote one PB, whence the great circle BA again falls below the zenith : in the latter, PA the nearest polar distance to the meridian in less than PB more remote, and the great circle BA falls above the Zenith. This difference affects the solution of the problem, as will presently appear.

The element in the solution most liable to error at the hands of a beginner is the polar angle BPA. But all difficulty under this head will be removed, if the student will remember, that,

(1) In the case of the sun, the Polar Angle BPA, is the interval of apparent time between the observations ; and, therefore, the equation of time must be applied to each of the two given chronometer times before the interval between them is taken.

(2) In the case of two stars, the Polar Angle BPA is the difference of their right ascensions, and, if there be an interval between the observations, this interval of mean time must be converted into sidereal time, and added to the right ascension of the first observed star, before the difference of the right ascensions is taken. If the observations of the stars are simultaneous, the Polar Angle is the difference of the right ascensions as they stand.

The work then proceeds as follows:—

Having given PA and PB the corrected polar distances, and ZA and ZB the Zenith distances, and BPA the Polar Angle, then,

(1) In the triangle BPA, with BP, PA and $\frac{1}{2}$ BPA, BA may be found, for

$$\text{Sin}^2\,\theta = \text{Sin PB . Sin PA . Cos}^2\,\tfrac{1}{2}\,\text{P}$$

$$\text{Sin}^2\,\frac{\text{AB}}{2} = \text{Sin}\left\{\frac{\text{PB} + \text{PA}}{2} + \theta\right\}\text{Sin}\left\{\frac{\text{PB} + \text{PA}}{2} - \theta\right\}$$

(2) In the same triangle $\frac{\text{PBA}}{2}$ may be found,

$$\text{Cos}\,\frac{\text{PBA}}{2} = \sqrt{\text{Sin S . Sin (S} - \text{PA) Cosec PB . Cosec BA}}$$

(3) In the triangle ZBA,

$$\text{Cos}\,\frac{\text{ZBA}}{2} = \sqrt{\text{Sin S . Sin (S} - \text{ZA) Cosec ZB . Cosec BA}}$$

(4) In figs. 1 and 2

$$\tfrac{1}{2}\,\text{PBA} - \tfrac{1}{2}\,\text{ZBA} = \tfrac{1}{2}\,\text{PBZ}$$

In figs. 3 and 4

$$\tfrac{1}{2}\,\text{PBA} + \tfrac{1}{2}\,\text{ZBA} = \tfrac{1}{2}\,\text{PBZ}$$

(5) In the triangle PBZ, with P B and BZ, and $\frac{1}{2}$ PBZ, the side PZ the colatitude may be found, for,

$$\text{Sin}^2\,\theta = \text{Sin PB . Sin BZ . Cos}^2\,\tfrac{1}{2}\,\text{PBZ}$$

$$\text{Sin}^2\,\frac{\text{PZ}}{2} = \text{Sin}\left\{\frac{\text{PB} + \text{BZ}}{2} + \theta\right\}\text{Sin}\left\{\frac{\text{PB} + \text{BZ}}{2} - \theta\right\}$$

Example I.—(Sun.)

January 10th, 1876, in lat by account 55° 20′ S, longitude 37° 30′ W, the following observations were taken to find the latitude.

Mean Times nearly.	Chon Times.	Obs Alts ⊙
$9^h\ 37^m$ A.M.	$1^h\ 50^m\ 30^s$	46° 54′ 10″
1 14 P.M.	5 27 30	55 5 55

Index Error + 1′ 30″; height of the eye 20 feet.

On December 20th, 1875, the chronometer was fast on Greenwich Mean Time, $1^h\ 30^m\ 17^s$, gaining daily 5^s.

Note.—(*a*) Find the approximate Greenwich dates from the mean times nearly; and the true Greenwich dates, by applying the error and rate of the chronometer to the chronometer times.

(*b*) With the true Greenwich dates find the declination of sun and equation of time.

(*c*) Correct the altitudes, and find the Zenith distances.

(*d*) The polar angle: found by applying the equation of time $\pm$ to mean time, to each chronometer time, and subtracting the first from the second,

(*e*) The computation by formulæ.

(1) Greenwich Date nearly.

	d	h	m	s
M.T. nearly	9	21	37	0
Longitude +		2	30	0
(1) G.D. nearly	10	0	7	0

Longitude.

°	′	
37	30 W	
	4	
6,0)15,0	0	
2 30	0	

(2) Greenwich Date nearly

	d	h	m	s
M.T. nearly	10	1	14	0
Longitude +		2	30	0
(2) G.D. nearly	10	3	44	0

(1) True Greenwich Date.

	h	m	s
Chron time	1	50	30
Rate −		1	45
	1	48	45
Error −	1	30	17
(1) T. G. Date 10^d	0	18	28

Rate.

5^s	per day
21	no. of days
6,0)10,5	
−1^m 45^s	

(2) True Greenwich Date.

	h	m	s
Chron times	5	27	30
Rate −		1	45
	5	25	45
Error −	1	30	17
(2) T.G. Date 10^d	3	55	28

Sun's Dec (Date 1).	Change.	Eq. of Time (Date 1).	Change.
° ′ ″	″	m s	s
S 22 1 0·5	−22·08	7 38·26	+1·016
6·6	·3	+·30	·3
22 0 53·9	6·624	7 38·56	·3048
90 0 0		—to Mean Time	
PA 67 59 6·1			

Sun's Dec (Date 2).	Change.	Eq. of Time (Date 2).	Change.
° ′ ″	″	m s	s
S 22 1 0·5	−22·08	7 38·26	+1·016
1 26·1	3·9	+3·96	3·9
21 59 34·4	19872	7 42·22	9144
90 0 0	6624		3048
BP 68 0 25·6	6,0)8,6·112		3·9624
	1′ 26″	— to Mean Time	

Corrected Altitudes.

1	° ′ ″	2	° ′ ″
Obs alt ⊙	46 54 10	Obs alt ⊙	55 5 55
IE	+ 1 30	IE	+ 1 30
	46 55 40		55 7 25
Dip	— 4 24	Dip	— 4 24
	46 51 16		55 3 1
Ref	— 53	Ref	— 40
	46 50 23		55 2 21
SD	+ 16 18	SD	+ 16 18
	47 6 41		55 18 39
Par^x	+ 6	Par^x	+ 5
(1) True alt	47 6 47	(2) True alt	55 18 44
	90 0 0		90 0 0
ZA	42 53 13	ZB	34 41 16

Polar Angle.

	h m s		h m s
(1) Chron time	1 50 30	(2) Chron time	5 27 30
(1) Eq time	— 7 38·6	(2) Eq time	— 7 42·2
	1 42 51·4		5 19 47·8

	h m s
	5 19 47.8
	1 42 51·4
2)	3 36 56·4
	1 48 28·2
	60
4)	108 28 12
½ P	27 7 3

N.B.—The first chronometer time is always subtracted from the second, which, if not large enough, must be increased by 12^h.

Figure.

Fig. 23.

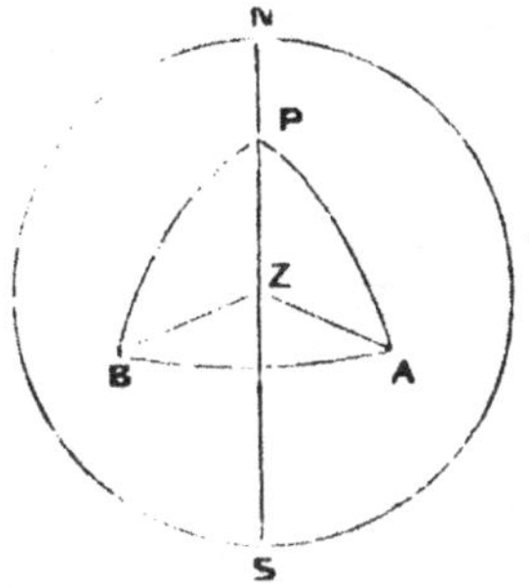

Here (1) the body at the first observation is east of the meridian, it being the sun, and the first mean time A.M.: and at the other west, the time being P.M.

And (2) since the polar distances PA, PB, are both greater than PZ the colatitude, the great circle BA falls below the Zenith.

The parts of the figures known, are

PA = 67° 59′ ZA = 42° 53′
PB = 68° 0′ ZB = 34° 41′ $\frac{1}{2}$P = 27° 7′

To find PZ

COMPUTATION (1)

$$\text{Sin}^2\,\theta = \text{Sin PA . Sin PB . Cos}^2\,\tfrac{1}{2}\text{P}$$

$$\text{Sin}^2\,\frac{\text{BA}}{2} = \text{Sin}\left\{\frac{\text{PA} + \text{PB}}{2} + \theta\right\}\text{Sin}\left\{\frac{\text{PB} + \text{PA}}{2} - \theta\right\}$$

	° ′		
$\frac{1}{2}$P =	27 7	cos	9·949429
			19·898858
PA =	67 59	sin	9·967115
PB =	68 0	sin	9·967166
	2)135 59		2)19·833139
$\frac{1}{2}$(PA + PB)	67 59	sin	9·916569
	θ 55 36		
Sum	123 35	sin	9·859962
Diff	12 23	sin	9·331329
			2)19·191291
$\frac{1}{2}$BA	23 12	sin	9·595645
	2		
BA =	46 24		

COMPUTATION (2)

$$\text{Cos}\,\frac{\text{PBA}}{2} = \sqrt{\text{Sin S . Sin (S — PA) Cosec PB . Cosec BA}}$$

	° ′		
PB	68 0	cosec	10·032834
BA	46 24	cosec	10·140158
PA	67 59		
	2)182 23		
S	91 11	sin	9·999907
PA	67 59		
(S — PA)	23 12	sin	9·595432
			2)19·768331
$\frac{1}{2}$PBA =	40 0	cos	9·884165

COMPUTATION (3)

$$\text{Cos}\,\frac{\text{ZBA}}{2} = \sqrt{\text{Sin S . Sin (S — ZA) Cosec ZB . Cosec BA}}$$

ZB	34° 41′	cosec	10·244857
BA	46 24	cosec	10·140158
ZA	42 53		
	2)123 58		
S	61 59	sin	9·945868
ZA	42 53		
(S — ZA)	19 6	sin	9·514837
			2)19·845720
½ZBA = 33° 9′		cos	9·922860

Then ½ PBA — ½ ZBA = ½ PBZ

½ PBA =	40° 0′
½ ZBA =	33 9
½ PBZ =	6 51

COMPUTATION (4)

$$\sin^2 \theta = \sin PB \,.\, \sin ZB \,.\, \cos^2 \tfrac{1}{2} PBZ$$

$$\sin^2 \frac{PZ}{2} = \text{in}\left\{\frac{PB + BZ}{2} + \theta\right\} \sin\left\{\frac{PB + BZ}{2} - \theta\right\}$$

½ PBZ	6° 51′	cos	9·996889
			2
			19·993778
PB	68 0	sin	9·967166
ZB	34 41	sin	9·755143
	2)102 41		2)19·716087
½ (PB + ZB)	51 20	sin	9·858043
θ	46 9		
Sum	97 29	sin	9·996285
Diff	5 11	sin	8·955894
			2)18·952179
½ PZ	17 25	sin	9·476089
	2		
Colatitude	34 50		
	90 0		
Latitude	55 10 S		

N.B.—Calculation to the nearest minute is sufficient for all practical purposes; but when a small arc occurs, the sine should be worked to seconds, as also the cosine of a large arc.

EXAMPLE II.—(TWO STARS.)

November 30th, 1876, in latitude by account 37° S, longitude 45° W, the following observations of stars were taken to find the latitude.

Mean Time nearly.	Chron Times.	Obs alts.
9h 24m P.M.	1h 12m 11s	α Leporis 41° 29′ 40″
9 58 P.M.	1 46 50	β Hydri 46 45 8

IE + 3′ 15″: height of the eye 14 feet.

Note.—(*a*) The right ascensions and declinations of the stars may be taken out at once for the given day, and the polar distances found, which name.

(*b*) Correct the altitudes, and find the zenith distances, which name.

(*c*) Sidereal interval, — *i.e.* the interval between the chronometer times converted into sidereal time.

(*d*) Polar angle = right ascension of first observed star + sidereal interval, and the difference between this, and right ascension of second star.

(*e*) Figure (explained below).

(*f*) Computation.

Right Asn α Leporis.	Decn α Leporis.	Right Asn β Hydri.	Decn β Hydri.
h m s	° ′ ″	h m s	° ′ ″
5 27 19·7	S 17 54 29	0 19 22·2	S 77 57 8
	90 0 0		90 0 0
	PL 72 5 31		PH 12 2 52

Sidereal Interval.

Chron times.		
h m s	m	m s
1 12 11	34 =	34 5·6
1 46 50	39s	39·1
Interval 34 39		34 44·7

Polar Angle.

	h m s
RA of 1st Star	5 27 19·7
Sidereal interval	+ 34 44·7
	6 2 4·4
RA of 2nd Star	0 19 22·2
	2)5 42 42·2
	2 51 21·1
	60
	4)171 21 6
½ P =	42 50 16

Corrected Zenith distances.

	° ′ ″		° ′ ″
1st alt	41 29 40	2nd alt	46 45 8
IE	+ 3 15	IE	+ 3 15
	41 32 55		46 48 23
Dip	— 3 41	Dip	— 3 41
	41 29 14		46 44 42
Ref	— 1 4	Ref	— 54
	41 28 10		46 43 48
	90 0 0		90 0 0
ZL	48 31 50	ZH	43 16 12

Figure.

1. To find whether the stars are east or west of the meridian.

Rule.—To the mean time at place add the sidereal time, this gives the right ascension of the meridian. Then, if the star's right ascension is greater than that of the meridian it is east, if less it is west.

Here,

Mean Time at place	9^h 24^m	9^h 58^m
Sidereal Time +	16 38	16 38
Right Asⁿ merⁿ	2 2	2 36

Now the Right Ascension of Leporis being 5^h 27^m, which is greater than 2^h 2^m, it is East of the meridian, while that of Hydri being 0^h 19^m, which is less than 2^h 36^m makes it West.

2. To find whether the great circle joining the places of the stars falls above or below the Zenith.

We have lat nearly 37°
90
∴ Colatitude PZ = 53

Polar dist Leporis PL 72°
„ Hydri PH 12

And, as the polar distance of Hydri differs more from the Colatitude than that of Leporis, the great circle falls between the Polar and Zenith distances, as in the following figure.

Fig. 24.

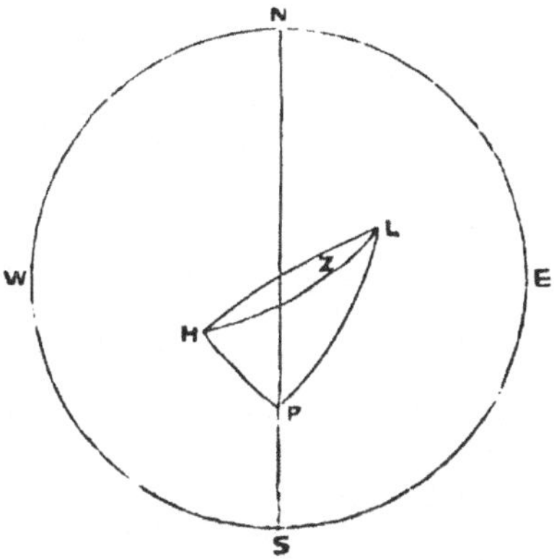

I. In the triangle HPL with PH, PL and ½P given, to find HL.

$$\text{Sin}^2\theta = \sin PH \,.\, \sin PL \,.\, \cos^2\tfrac{1}{2}P$$

$$\text{Sin}^2\frac{HL}{2} = \sin\left\{\frac{PH + PL}{2} + \theta\right\}\sin\left\{\frac{PH + PL}{2} - \theta\right\}$$

½P =	42°	50′	cos	9·865302
				2
				19·730604
PH	12	3	sin	9·319658
PL	72	5	sin	9·978411
	2)84	8		2)19·028673
½(PH + PL	42	4	sin	9·514336
θ	19	4		
Sum	61	8	sin	9·942378
Diff	23	0	sin	9·591878
				2)19·534256
½HL	35°	48′	sin	9·767128
		2		
HL	71	36		

II. In the triangle PLH, with PL, LH, and PH, to find ½PLH.

$$\text{Cos } \tfrac{1}{2}\text{PLH} = \sqrt{\text{Sin S.Sin (S} - \text{PH) Cosec PL Cosec LH}}$$

PL	72°	5′	cosec	10·021589
LH	71	36	cosec	10·022791
PH	12	3		
	2)155	44		
S	77	52	sin	9·990188
PH	12	3		
(S — PH)	65	49	sin	9·960109
				2)19.994677
½PLH	6°	20′	cos	9·997338

III. In the triangle ZLH, with ZL, LH, and ZH, to find ½ZLH.

$$\text{Cos } \tfrac{1}{2}\text{ZLH} = \sqrt{\text{Sin S.Sin (S} - \text{ZH) Cosec ZL Cosec LH}}$$

ZL	48°	32′	cosec	10·125320
LH	71	36	cosec	10·022791
ZH	43	16		
	2)163	24		
S	81	42	sin	9·995427
ZH	43	16		
(S — ZH)	38	26	sin	9·793514
				2)19·937052
½ZLH	21°	33′	cos	9·968526

IV. Then $\frac{1}{2}$ PLH + $\frac{1}{2}$ ZLH = $\frac{1}{2}$ PLZ.

$\frac{1}{2}$PLH	6°	20′
$\frac{1}{2}$ZLH	21	33
$\frac{1}{2}$PLZ	27	53

V. In the triangle PLZ, with PL, ZL, and $\frac{1}{2}$PLZ, to find PZ, the colatitude.

$$\text{Sin}^2\theta = \text{Sin PL.Sin ZL Cos}^2\tfrac{1}{2}\text{PLZ}$$

$$\text{Sin}^2\frac{\text{PZ}}{2} = \text{Sin}\left\{\frac{\text{PL}+\text{ZL}}{2}+\theta\right\}\text{Sin}\left\{\frac{\text{PL}+\text{ZL}}{2}-\theta\right\}$$

	°	′		
$\frac{1}{2}$PLZ	27	53	cos	9·946404
				2
				19·892808
PL	72	5	sin	9·978411
ZL	48	32	sin	9·874680
	2)120	37		2)19·745899
$\frac{1}{2}$(PL + ZL)	60	18	sin	9·872949
θ	48	16		
Sum	108	34	sin	9·976787
Diff	12	2	sin	9·319066
				2)19·295853
	26	23	sin	9·647926
		2		
Colat	52	46		
	90			
Lat	37	14 S		

EXAMPLES FOR EXERCISE.

1. March 8th 1876, in lat by account 54° 20′ N, long 27° 20′ W, the following observations were taken to find the latitude.

Mean Times nearly	Chron Times	Obs Alts ☉
$8^h\ 27^m$ A.M.	$9^h\ 51^m\ 45^s$	14° 38′ 45″
2 24 P.M.	3 48 45	24 35 37

Index error — 1′ 19″: height of the eye 19 feet.

On February 10th, at noon, the chron was slow on Greenwich Mean Time, $24^m\ 30^s$, losing daily 2^s.

Ans. Lat 54° 30′ N.

2. November 24th, at $8^h\ 10^m$ P.M., mean time nearly in lat 40° N, long 20° 20′ W., the following altitudes of stars were taken.

Obs alt α Arietis	Obs alt β Tauri
63° 56′ 37″	30° 28′ 31″
IE + 5 10	IE — 2 55

Height of the eye 21 feet.

Ans. Lat 40° 2′ N.

3. April 15th. at 10^h P.M., mean time nearly in lat by account 51° 30′ S. long 36° 20′ W.

Mean Time nearly.	Chron Times.	Obs Alts.
9^h 55^m P.M.	4^h 31^m 12^s	Sirius 23° 21′ 15″
10 0 P.M.	4 35 55	Regulus 22 41 10

Index error — 1′ 11″: height of the eye 18 feet.

Ans. Lat 51° 34′ S.

ELEMENTS FOR PRECEDING EXAMPLES FROM NAUTICAL ALMANAC, 1876.

1. March 7th, dec S 5° 1′ 15″ change 58″ 40 — ; Equation Time 11^m $5^s.7$ Change — .621^s — to Mean Time : Semi diam 16′ 8″.

2. RA of Arietis 2^h 0^m 26^s: Dec^n N 22° 54′
RA of Tauri 5 18 44 Dec^n N 28 30
Sidereal time 16^h 12^m 17^s.

3. RA of Sirius 6^h 39^m 50^s: Dec^n S 16° 33′
RA of Regulus 10 1 58 Dec^n N 12 33
Sidereal time 1^h 33^m 5^s.

LONGITUDE.

(11.) LONGITUDE BY CHRONOMETER.

This is, perhaps, the most valuable of all problems in Nautical Astronomy, and one in general use at sea. The chronometer is an instrument which is supposed to record the true time at Greenwich at any instant it may be required ; *i.e.*, to the time shown by it, the known error and rate are applied, from which the Greenwich mean time is obtained, (see article on chronometers.) Questions are often proposed, in which the error of the chronometer for mean or apparent time at place is given : this *in practice* would be unmeaning. The use of the chronometer is not to show time at place, but the Greenwich time at any instant required. The former is reckoned by an entirely separate and independent instrument called the "Hack watch." Let it be remembered, then, that chronometers show Greenwich mean time, the object for which they are made.

This method may be said to consist of two parts, viz—

1. From an approximate time to find the true mean time at place.
2. Comparing this with the Greenwich date obtained from the chronometer, find the longitude, remembering the following little rule,—

> Greenwich date least,
> Longitude East.
> Greenwich date best,
> Longitude West.

This may be illustrated as follows.

EXAMPLE 1.

Let the date at place as deduced from the work be $5^h\ 27^m\ 30^s{\cdot}5$, and the Greenwich date $7^h\ 29^m\ 10^s{\cdot}5$, then if the given longitude be west, the result must be,

True time at place	5^h	27^m	$30^s{\cdot}5$
Greenwich date	7	29	10.5
Difference	2	1	40·0
	60		
	4)121		40·0
Longitude	30°	25·0″	West

EXAMPLE 2.

Let the date at place be $8^h\ 21^m\ 40^s$, and the Greenwich date be $2^h\ 40^m\ 10^s$, then, the longitude being East, the result is,

True time at place	8^h	21^m	40^s
Greenwich date	2	40	10
Difference	5	41	30
	60		
	4)341		30
Longitude	85°	22′	30″ East

EXAMPLE 3.

Suppose now the date deduced from work to be $17^h\ 20^m\ 10^s$, and the Greenwich date $1^h\ 40^m\ 30^s$, the longitude being west. In this case, (remembering the foregoing rule), 24^h must be added to the Greenwich date, to make it "best," thus

True time at place	17^h	20^m	10^s
Greenwich date	25	40	30
Difference	8	20	20
	60		
	4)500		20
Longitude	125°	5′	0″ West

EXAMPLE 4.

Conversely, suppose the approximate longitude to be East, the true mean time at place being $5^h\ 21^m\ 10^s{\cdot}5$, and the Greenwich date $21^h\ 40^m\ 7^s$, then 24^h must be added to the former to make the longitude **east**, or Greenwich date **least**, hence,

True time at place	29^h	21^m	$10^s{\cdot}5$
Greenwich date	21	40	7
Difference	7	41	3·5
	60		
	4)461		3·30
Longitude	115°	15′	52″ East

The two latter cases are those in which the student usually finds most difficulty, but which need not be difficult, if the rules be remembered.

EXAMPLE 1. (SUN).

On March 3rd. 1876, at about $9^h\ 37^m$ A.M., mean time nearly, in lat 48° 10′ N, long by account 2° 40′ E, when a chronometer showed $8^h\ 46^m\ 10^s$, the obs alt ☉ was 25° 7′ 0″, IE + 1′ 25″ : Height of the eye 13 feet: required the longitude.

On February 20th, the chronometer was slow on Greenwich mean time $39^m\ 44^s$, losing daily 3 seconds.

Note.—(*a*) Find the approximate Greenwich date from the mean time at place by applying the longitude in time.

(*b*) Apply the error and rate of the chronometer to the time it showed when the observation was taken. If **slow**, add these: if **fast, subtract** them. Then if the **approximate** Greenwich date exceeds 12^h, that number must be added to this date to make the hours of the two agree nearly. When this is done, the former date may be rejected, it being used as a check upon the chronometer date.

(*c*) For this, take out the Sun's declination, and the equation of time.

(*d*) Correct the altitude.

(*e*) Find the hour angle from the formula,

$$\text{Sin}\ \tfrac{1}{2}\ h = \sqrt{\text{Sec}\ l\ \text{Cosec}\ p\,.\,\text{Cos S}\,.\,\text{Sin}\ (\text{S}-a)}$$

(*f*) Reduce this to time: then if the observation be in the morning, the sun is **east** of the meridian and the hour angle must be subtracted from 24^h, to obtain the westerly meridian distance, or true **apparent** time, at place. To this apply the equation of time $\pm$ to apparent time, the result being the true mean time at place. Take the difference between this, and the chronometer Greenwich date, as explained above, and the result is the longitude.

Greenwich Date nearly.		Longitude.	True Greenwich Date.		Rate.
	d h m s	° ′		h m s	s
Date at place	2 21 37 0	2 40 E	Chron Time	8 46 10	3 pr day
Longitude	— 10 40	4	Error	+ 39 44	× 11 days
G D nearly	2 21 26 20	10 40		9 25 54	= 33
			Rate	+ 33	
				9 26 27	
			Add	12 0 0	
			T G D Mar 2ᵈ	21 26 27	

	Sun's Declination.	Change.	Equation of Time.	Change.
	′ ″	″	m s	s
	S 6 57 9·3	— 57·41	12—14·84	— ·528
	— 20 28·6	21·4	—11·30	21·4
	6 36 40·7	22964	12— 3·54	2112
	90 0 0	5741	+ App. T.	528
Polar dist	96 36 41	11482		1056
		6,0)122,8·574		11·2992
		20′ 28″ 6		

Corrected Altitude.

	° ′ ″
Obs alt ⊙	25 7 0
IE	+ 1 25
	25 8 25
Dip	— 3 33
	25 4 52
Ref	— 2 2
	25 2 50
SD	+ 16 10
	25 19 0
Parallax	+ 8
True alt (a)	25 19 8

Hour Angle.

$$\sin \tfrac{1}{2}h = \sqrt{\sec \text{lat} \operatorname{cosec} p \,.\, \cos S \,.\, \sin(S-a)}$$

	° ′ ″		
a	25 19 8		
l	48 10 0	sec	10·175896
p	96 36 41	cosec	10·002898
	2)170 5 49		
S	85 2 54	cos	8·936088
a	25 19 8		
(S−a)	59 43 46	sin	9·936340
			2)19·051222
	19° 35′ 58″	sin	9·525611
	2		
	39 11 56		
	4		
	6,0)15,6 47 44		
Hour Angle	2 36 47·7		

Longitude.

	h	m	s	
Hour Angle	2	36	47·7	
Subtract from	24	0	0	(time A.M.)
True app : Time at place	21	23	12·3	
Equation of Time	+	12	3·5	
True Mean Time at place	21	35	15·8	
Greenwich Date	21	26	27·0	
	0	8	48·8	
	60			
4)	8	48	48	
Longitude	2°	12′	12″	East.

EXAMPLE II.

April 11th, 1876, at about 3^h 37^m P.M., mean time nearly, in latitude 44° 30′ N, longitude by account 17° 40′ W, when a chronometer showed 6^h 2^m 29^s, the obs alt ☉ was 31° 19′ 20″, IE + 1′ 20″: height of the eye 16 feet: required the longitude.

On March 11th, the chronometer was fast on Greenwich mean time 1^h 13^m 12^s, gaining daily 3^s.

Greenwich Date nearly.

	d	h	m	s
Date at place	11	3	37	0
Longitude	+	1	10	40
Greenwich Date nearly	11	4	47	40

Longitude.

°	′	
17	40 W	
	4	
6,0)7,0	40	
1	10	40

True Greenwich Date.

	h	m	s
Chron showed	6	2	29
Error	1	13	12
	4	49	17
Rate	—	1	33
Time Green^h^ date at pl 11^d^	4	47	44

Rate.

3^s^ per day
× 31
93^s^

Sun's Declination.	Change.	Equation of Time.	Change.
N 8° 33′ 2″	+ 54″·84	0^m^ 55^s^·20	—^s^·669
+ 4 23	4·8	3·21	4·8
8 37 25	43872	0 51·99	5352
90 0 0	21936	+ App Time.	2676
p 81 22 35	6,0)26,3·232		3·2112
	+ 4 23·2		

Corrected Altitude.

	° ′ ″
Obs alt $\overline{\odot}$	31 19 20
IE	+ 1 20
	31 20 40
Dip	— 3 56
	31 16 44
Ref	— 1 33
	31 15 11
SD	— 15 59
	30 59 12
Par^x	+ 8
(a)	30 59 20

Hour Angle.

$$\sin \tfrac{1}{2} h = \sqrt{\sec l \operatorname{cosec} p . \cos S . \sin (S-a)}$$

	° ′ ″		
a	30 59 20		
l	44 30 0	sec	10·146758
p	81 22 35	cosec	10·004938
	2)156 51 55		
S	78 25 57	cos	9·302162
a	30 59 20		
(S—a)	47 26 37	sin	9·867238
			2)19·321096
	27 14 11	sin	9·660548
	2		
	54 28 22		
	4		
	6,0)21,7 53 28		
Hour angle	3 37 53·5		

Longitude.

	h m s	
Hour angle (app time at place)	3 37 53·5	(time P.M.)
Equation of time	+ 51·9	
True mean time at place	3 38 45·4	
Greenwich date	4 47 44·0	
	1 8 58·6	
	60	
	4)68 58 36	
Longitude	17° 14′ 39″	West

EXAMPLE III. (With a Star).

In this, the work does not differ much from the preceding, until the hour angle is found, when (if not stated previously) it must be calculated whether the star is east or west of the meridian at the time of observation, (see fig. in double altitude by direct spherics): if the former, subtract from 24^h to obtain the westerly hour angle: then

Star's westerly hour ∠ + star's RA — mean sun's RA = true mean time at place.

N.B.—The mean sun's RA is **always subtracted from** the sum of the other two elements, which, if not large enough, must be **increased by 24^h.**

Then the remainder as before.

On May 10th, 1876, at about 5^h 42^m P.M., mean time nearly, in lat 12° 42′

S, longitude by account 6° 20′ W, when a chronometer showed 7^h 32^m 49^s, the obs alt of Regulus was 59° 50′ 20″, IE + 1′ 15″, height of the eye 13 feet: required the longitude.

April 10th, the chronometer was fast on Greenwich mean time 1^h 21^m 47^s, gaining daily 5 seconds.

Greenwich Date nearly.		Longitude.	True Greenwich Date.		Rate.
	d h m s	° ′		h m s	s
Date at place	10 5 42 0	6 20 W	Chron showed	7 32 49	5 per day
Longitude	+ 25 20	4	Error —	1 21 47	× 30 days
G. D. nearly	10 6 7 20	25 20		6 11 2	150^s
			Rate	2 30	
			T. G. date, May	10^d 6 8 32	

Star's Right As^n.	Star's Polar Distance.		Mean Sun's Right As^n.	
h m s	° ′ ″			h m s
10 1 47·4	N 12 34 18		Sidereal time	3 14 30·3
	90 0 0		Accelerat^n for 6^h =	59·1
	p 102 34 18		Table „ 8^m =	1·3
			Naut^l Alm^c „ 32^s =	·1
				3 15 30·8

Corrected Altitude.

	° ′ ″
Obs alt	59 50 20
IE	+ 1 15
	59 51 35
Dip	— 3 33
	59 48 2
Ref	— 34
(a)	59 47 28

Hour Angle.

$$\text{Sin } \tfrac{1}{2} h = \sqrt{\text{Sec } l \text{ cosec } p . \cos S . \sin (S - a)}$$

	° ′ ″		
a	59 47 28		
l	12 42 0	sec	10·010757
p	102 34 18	cosec	10·010539
	2)175 3 46		
S	87 31 53	cos	8·634194
a	59 47 28		
(S—a)	27 44 25	sin	9·667886
			2)18·323376
	8 20 37	sin	9·161688
	2		
	16 41 14		
	4		
	6,0)6,6 44 56		
Hour angle	1 6 44·9		

To find whether the Star is East or West of the Meridian.

Rule. Mean Time place + Mean Sun's R*a* = R*a* of mer.

Then if Star's *Ra* be greater than *Ra* of meridian it is **East.**

„ „ less „ „ **West.**

Here	Mean time place	5^h 42^m	Then *Ra* of Star is 10^h:
	Mean Sun's *Ra*	3 15	∴ Star is East of meridian
	∴ *Ra* of meridian	8 57	

Longitude.

	h	m	s
Easterly Hour Angle	1	6	44·9
Subtract from	24	0	0
West Hour Angle of Star	22	53	15·1
Add Star's Right as^n	10	1	47·4
Ra of meridian	32	55	2·5
Subtract Mean Sun's *Ra*	3	15	30·8
	29	39	31·7
Reject	24	0	0
True Mean Time at place	5	39	31·7
Greenwich Date	6	8	32·0
	0	29	0·3
	60		
	4)29	0	18
Longitude	7°	15′	4″·5 West.

N.B. The Right Ascensions being reckoned round to 24^h, such a case as follows may occur,—

Suppose *Ra* of meridian = 1^h

and *Ra* of Star = 23

Then as there cannot be 22^h difference of *Ra* when the object is above the horizon, the *Ra* of meridian must be called 25^h, from which it will appear, that the Star is **West** of the meridian, and **not East.**

EXAMPLES FOR EXERCISE.

1. May 30th, 1876, at 8^h 34^m A.M., mean time nearly in latitude 24° 30′ N, longitude by account 25° 25′ E, when a chronometer showed 5^h 26^m 2^s, the obs alt $\underline{\odot}$ was 43° 29′ 0″. Index Error — 1′ 17″; height of the eye 20 feet: required the longitude. On May 1st, the chronometer was slow on Greenwich Mean Time, 1^h 24^m 20^s losing daily 5 seconds.

Ans. Longitude 25° 30′ 0″ E

2. June 8th, at about 2^h 37^m P.M., mean time nearly, in lat 51° 10′ N, longitude by account 12° 40′ W, when a chronometer showed 2^h 37^m 44^s, the obs alt $\underline{\odot}$ was 48° 9′ 0″, IE + 1′ 15″: height of the eye 20 feet: required the longitude.

May 20th, the chronometer was slow on GMT, 49^m 47^s, losing daily 3 seconds.

Ans. Long 12° 36′ 40″ W.

3. July 20th, at about $7^h\ 35^m$ A.M., mean time nearly, in lat 59° 30′ N, longitude by account 23° 40′ W, when a chronometer showed $8^h\ 38^m\ 21^s$, the obs alt $\underline{\odot}$ was 29° 14′ 30″, IE — 1′ 30″ : height of the eye 24 feet: required the longitude.

July 1st, the chronometer was slow on GMT, $31^m\ 24^s$, losing daily 3 seconds.

Ans. Long 22° 55′ 30″ W.

4. March 17th, at about $8^h\ 40^m$ P.M., mean time, in lat 59° 40′ N, longitude by account 17° 20′ W, when a chronometer showed $11^h\ 28^m\ 45^s$, the obs alt *a* Orionis 30° 28′ 10″, IE — 1′ 10″: height of the eye 25 feet: required the longitude.

March 1st, the chronometer was fast on GMT, $1^h\ 37^m\ 27^s$, gaining daily 3 seconds.

Ans. Long 18° 14′ 30″ W.

5. April 18th, at about $6^h\ 30^m$ P.M., mean time in lat 48° 17′ N, longitude by account 16° 25′ E, when a chronometer showed $3^h\ 44^m\ 7^s$, the obs alt of *a* Hydræ was 31° 51′ 40″, IE + 1″·10 : height of the eye 16 feet: required the longitude.

April 2nd, the chronometer was slow on GMT, $1^h\ 39^m\ 27^s$, losing daily 3^s·5.

Ans. Long 15° 46′ 34″ E.

ELEMENTS FOR PRECEDING EXAMPLES FROM NAUTICAL ALMANAC, 1876.

1. Sun's dec N 21° 43′ 22″ : change + 22″·68 ;
 Equation of Time $2^m\ 48^s$·36 ; change — ·331^s, — to app T.
 Sun's semi diam 15′ 48″.
2. Sun's dec N 22° 54′ 44″ : change + 12″·92 ;
 Equation of Time $1^m\ 10^s$·4 ; change — ·475^s, — to app T.
 Sun's semi diam 15′ 47″.
3. Sun's dec N 20° 45′ 35″ : change — 27″·76 ;
 Equation of Time $6^m\ 1^s$·53 ; change + ·162^s, + to app T.
 Sun's semi diam 15′ 47″.
4. Sidereal Time $23^h\ 41^m\ 36^s$·4 : Star's RA $5^h\ 48^m\ 28^s$·1 :
 Dec 7° 22′ 59″ N
5. Sidereal Time $1^h\ 47^m\ 46^s$·1 : Star's RA $9^h\ 21^m\ 30^s$·8 ;
 Dec 8° 7′ 22″ S.

(12). TO FIND THE LONGITUDE FROM A LUNAR OBSERVATION.

As the moon is the nearest heavenly body to the earth, her motion through the celestial sphere is more rapid than that of any other object, and her distance from the sun or stars as apparent to the observer, undergoes continuous change.

Her distance from the sun and certain stars, is calculated in the Nautical Almanac for every three hours at Greenwich: hence, from an observed distance at any place on the earth's surface, the Greenwich date may be found by the medium of a table of proportional logarithms. Thus, in the Almanac opposite each distance, or between two successive lunar distances is inserted a proportional logarithm of their difference. By subtracting this log from the proportional log of the difference between the true distance from observation, and that preceding in the Almanac, gives the proportional log of a number of hours, minutes and seconds to be added to the hours over the distance in the Almanac to find the Greenwich time.

But the distance between the moon and any object, as seen from the surface of the earth, is considerably different from the corresponding distance which would be seen by an observer at the earth's centre. If it were not so, the work of a lunar would be simple enough ; it would be sufficient merely to observe the distance with a sextant, correct it for Index Error, then compare the distance with that in the Almanac, and thus obtain the Greenwich date.

The moon's true altitude, however, through correction for parallax, or reduction to what it would be if taken from the earth's centre, is always larger than her apparent altitude, inasmuch as her parallax predominates in amount largely over her refraction. The reverse is the case in the altitude of a star or of the sun, the refraction always is the larger. To reduce the observed distance, then, as seen from the surface, to what it would be if seen from the centre of the earth, is to "clear the distance," *i.e.*, to free it from the effects of parallax and refraction.

The chief value of the lunar method, is, that, when the altitudes are observed at the same time as the distance, the exact Greenwich date is deduced from a very approximate one ; the elements used, *e.g.*, the change in the moon's SD, and horizontal parallax, being but little affected by, say an interval of an hour or half-an-hour.

When, however, the altitudes are not observed, but only the distance, the true mean time at place of observation must be precisely determined, from which the altitudes are computed. This method is one very rarely adopted at sea, both on account of the length of the calculation, and also of the minuteness required in the error of chronometer and distance ; and that this should not be resorted to is not surprising, when with three simultaneous, tolerably exact observations, and a very rough time at ship, an exact and accurate result should be determined; or one observer alone, may, by changing about with distances and altitudes, reduce the former to a simultaneous condition with the latter.

We shall now, before exhibiting the problem at length, dwell upon the two important parts of it, in order, viz., (1) from an observed distance to find the true, and (2) from a true distance to find the Greenwich date by means of the Almanac distance and table of proportional logs. (See page 75).

EXAMPLES.

Given.	Appt dist between Sun and Moon	$(d) =$	80°	24′	56″
	Appt alt of Sun	$(s) =$	15	44	7
	Appt alt of Moon	$(m) =$	51	53	47
	True alt of Sun	$(S) =$	15	40	54
	True alt of Moon	$(M) =$	52	28	22
	To find the true distance .	(D)			

Some time may be saved by rejecting the seconds, as follows :— if the minutes of (d), (s), (m), add up evenly, take all the seconds off: if unevenly, add on sufficient seconds to (d) to make another minute.

Reject the same number of seconds from the true alts (S). (M) as are taken from (s), (m), add on to (D), when obtained, the seconds taken from (d) : Or the reverse: Thus :—

$$\sin^2\theta = \cos M \cos S . \cos x . \cos(x - d) \sec m \sec s.$$

$$\sin^2\frac{D}{2} = \cos\left(\frac{M+S}{2}+\theta\right)\cos\left(\frac{M+S}{2}-\theta\right)$$

(d) 80° 24′ 56″ (add 4″) =	80° 25′		
(s) 15 44 7 (reject 7) =	15 44	sec	10·016584
(m) 51 53 47 (reject 47) =	51 53	sec	10·209529
	2)148 2		
x	74 1	cos	9·439897
d	80 25		
$(d-x)$	6 24	cos	9·997285

(S) 15° 40′ 54″ (reject 7″) =	15° 40′ 47″	cos	9·983530
(M) 52 28 22 (reject 47) =	52 27 35	cos	9·784845
	2)68 8 22		2)19·431670
½(M+S)	34 4 11	sin	9·715835
θ	31 19 8		
[½(M+S)+θ]	65 23 19	cos	9·619575
[½(M+S)—θ]	2 45 3	cos	9·999500
			2)19·619075
½D	40 9 48	sin	9·809537
			2
			80 19 36
		(Secs + to d)	— 4
		(D)	80 19 32

EXAMPLES FOR EXERCISE.

(1).

Given (*d*)	38° 10′ 34″	(S)	49° 58′ 34″
„ (*s*)	49 59 22	(M)	66 43 22
„ (*m*)	66 19 58		
Find (D)			

Ans. 38° 7′ 0″

(2).

Given (*d*)	89° 35′ 46″	(S)	30° 23′ 44″
„ (*s*)	30 25 21	(M)	40 7 48
„ (*m*)	39 25 56		
Find (D)			

Ans. 89° 9′ 44″

(3).

Given (*d*)	53° 52′ 2″	(S)	39° 57′ 32″
„ (*s*)	39 58 34	(M)	63 18 12
„ (*m*)	62 51 27		
Find (D)			

Ans. 53° 44′ 25″

Greenwich Date from True Distance.

Note.—(*a*) Look in the Nautical Almanac, on the given day, for the distance under the next less number of hours to that in the approximate Greenwich Date, and write the proportional log opposite.

(*b*) The difference between the true distance and the almanac distance.

(*c*) Proportional log of difference.

(*d*) Difference of the two proportional logs.

(*e*) Number of hours &c., opposite this from same table, add on the number of hours over distance in almanac.

(*f*) Correction for second differences.

EXAMPLE.

February 16th, 1876, in longitude 175° 40′ E, about 8ʰ 30ᵐ A.M., mean time nearly, the true distance between sun and moon was 104° 47′ 16″, find the correct Greenwich Date.

Greenwich Date nearly.	d h m s	Longitude.
February	15 20 30 0	175° 40′ E
	— 11 42 40	4
	15 8 47 20	6,0)70,2 40
		11 42 40

		° ′ ″		Prop^l logs.
	(D) =	104 47 16		——
	Feb 15th dist sun at VI. =	106 5 28	. .	3308
	Diff =	1 18 12	. .	3621
		$2^h\ 47^m\ 29^s$	= diff	313
	add	VI		
Correction	Interval = $2^h\ 47^m$ }	15^d 8 47 29		
	Diff of logs = 8 }	. . . + 4		
	True Greenwich Date	8 47 33		

EXAMPLES FOR EXERCISE.

1. February 6th, at about $7^h\ 30^m$ P.M., in long 102° 20′ E, the true distance between *a* Arietis and the moon was 56° 54′ 18″: find the Greenwich date.

Ans. February $6^d\ 0^h\ 40^m\ 28^s$.

2. October 22nd, at about $2^h\ 50^m$ P.M., in long 12° 20′ E, the true distance between sun and moon was 60° 13′ 0″: find the Greenwich date.

Ans. October $22^d\ 2^h\ 0^m\ 31^s$.

3. July 10th, at about $2^h\ 40^m$ A.M., in long 6° 20′ E, the true distance between *a* Pegasi and moon was 33° 19′ 40″: find the Greenwich date.

Ans. July $9^d\ 14^h\ 13^m\ 34^s$.

Lunar Observations. (Sun).

1. July 18th 1876, at about $9^h\ 30^m$ A.M., mean time nearly, in lat 57° 30′ N, long by account 110° 35′ E, the following lunar was taken.

Obs alt ☉	Obs alt ☾	Obs dist.
43° 42′ 0″	58° 12′ 0″	43° 55′ 0″
IE — 1 13	IE — 1 30	IE — 1 30

Height of the eye 12 feet; find the longitude.

Note.—(*a*) Find the approximate Greenwich date.

(*b*) Take out the ☾'s SD and Horizontal Parallax, which correct for change, and for augmentation and reduction from table.

(*c*) Correct the altitudes, taking care to apply semi-diameter before refraction.

(*d*) Apply the SD of both objects, as also the index error to (*d*); and, as the moon's bright edge is always turned to the sun, the distance observed is that of nearest limbs; therefore, the SD^s must both be added to obtain distance of centres (*d*).

(*e*) Clear the distance and find Greenwich date.

(*f*) Find the true mean time at place, if not known, and work the longitude as by chronometer.

Greenwich Date nearly.	Longitude.
d h m s	° ′
17 21 30 0	110 35 E
— 7 22 20	4
17 14 7 40	6,0)44,2 20
	7 22·20

(·SD.	Change.	(·HP.	Change.
′ ″	″	′ ″	″
16 27·5	+ 5·2	60 18·0	+ 19·1
+ ·9	2·1	+ 3·4	2·1
16 28·4	52	60 21·4	191
Aug + 15·2	104	Red — 8·2	382
16 43·6	12)10·92	60 13·2	12)40,11
	·9	60	3·4
		3613″ HP.	

Altitudes.

(*s*) and (S)		(*m*) and (M)	
	° ′ ″		° ′ ″
Obs alt ☉	43 42 0	☾	58 12 0
IE	— 1 13	IE	— 1 30
	43 40 47		58 10 30
Dip	— 3 25	Dip	— 3 25
	43 37 22		58 7 5
SD	+ 15 46	SD	+ 16 44
(*s*)	43 53 8	(*m*)	58 23 49
Ref	— 59	Ref	— 35
	43 52 9		58 23 14
Parallax	+ 6	Par^x	+ 31 34
(S)	43 52 15	(M)	58 54 48

(·Parallax in Altitude.

Log HP 3613 = 3·557868
Log cos alt = 9·719525
6,0)189″,4 = 3·277393
+ 31·34

(*d*)

	° ′ ″
Obs dist	43 55 0
IE	— 1 30
	43 53 30
☉·SD +	15 46
(·SD +	16 44
(*d*)	44 26 0

True Distance (D).

(d)	44° 26′ 0″ ...	=	44° 26′ 0″		
(s)	43 53 8 (−8″)	=	43 53 0	sec	10·142214
(m)	58 23 49 (−49)	=	58 23 0	sec	10·280475
			2)146 42		
		x	73 21	cos	9·457162
		d	44 26		
		(x—d)	28 55	cos	9·942169
(S)	43 52 15 (—8″)	=	43 52 7	cos	9·857894
(M)	58 54 48 (−49)	=	58 53 59	cos	9·713098
			2)102 46 6		2)19·393012
		½ (M + S)	51 23 8	sin	9·696506
		θ	29 48 47		
		[½ (M + S) + θ]	81 11 50	cos	9·184787
		[½ (M + S) — θ]	21 34 16	cos	9·968466
					2)19·153253
		½D	22° 9′ 48″	Sin	9·576626
			2		
		True dist D =	44 19 36		

True Greenwich Date.

	° ′ ″		Prop[l] log.
D =	44 19 36		
Dist at XII. =	45 32 35	—	2440
Diff	1 12 59	—	3920
	2[h] 8[m] 1[s]	=	1480
Add	12		
	17[d] 14 8 1		
Cor for second diff[s]	— 4		
	17 14 7 57		Greenwich Date

Longitude.

Sun's Declination.	Change.	Equation of Time.	Change.
° ′ ″	″	m s	s
N 21 7 4	— 25·99	5—52·69	+ ·206
— 6 6	14· 1	+ 2·90	14·1
21 0 58	2599	5—55·59	206
90 0 0	10396	+ A.T.	824
p 68 59 2	2599		206
	6,0)366,459		2·9046
	6′ 6″·4		

To find the Longitude.

	° ′ ″		
Altitude (*a*)	43 52 15		
Latitude (*l*)	57 30 0	sec	10·269784
Polar dist (*p*)	68 59 2	cosec	10·029896
	2)170 21 17		
	S 85 10 38	cos	8·924661
	a 43 52 15		
(S—*a*)	41 18 23	sin	9·819600
			2)19·043941
½*h*	19 25 45	sin	9·521970
	2		
	38 51 30		
	4		
	6,0)15,5 26 0		
Easterly Hour ∠	2 35 26		
	24 0 0		
True app Time at place	21 24 34		
Equation Time	+ 5 55·6		
True mean time at place	21 30 29·6		
Greenwich Date	14 7 57·0		
	7 22 32·6		
	60		
	4)442 32 36		
Longitude	110 38 9	East	

Lunar Observations. (Star).

April 10th, 1876, at about 2^h 30^m A.M., mean time in lat 27° 30′ N, long by account 86° 10′ E, the following lunar was taken, to determine the longitude.

Obs alt *a* Aquilœ (Altair)	Obs alt ☾	Obs dist FL
30° 29′ 0	39° 14′ 0″	89° 49′ 50″
IE + 1 10	IE + 1 23	IE + 1 18

Height of the eye 24 feet.

Greenwich Date nearly.	Longitude.
d h m s	° ′
9 14 30 0	86 10 E
5 44 40	4
9 8 45 20	6,0)34,4 40
	5 44 40

☾ SD	Change.	☾ HP.	Change.
′ ″	″	′ ″	″
15 16·2	−4·3	55 56·6	— 15·6
—3·1	8·7	—11·3	8·7
15 13·1	301	55 45·3	1092
Aug + 9·3	344	Redⁿ — 2·5	1248
15 22·4	12)37·41	55 42·8	12)135·72
	—3·1	60	—11·3
		HP 3343″	

Altitudes.

	° ′ ″		° ′ ″
Obs alt star	30 29 0	Obs alt ☾	39 14 0
IE +	1 10	IE +	1 23
	30 30 10		39 15 23
Dip —	4 49	Dip —	4 49
(*s*)	30 25 21		39 10 34
Ref —	1 37	SD +	15 22
(S)	30 23 44	(*m*)	39 25 56
		Ref —	1 9
			39 24 47
		Parˣ +	43 2
		(M)	40 7 49

☾ Parallax in alt.

Log HP 3343″ = 3·524136
Log cos alt = 9·887926
6,0)258″,2 = 3·412062
+43′ 2″
(*d*)

	° ′ ″
Obs dist	89 49 50
IE +	1 18
	89 51 8
☾ SD —	15 22
(*d*)	89 35 46

The (☾ SD) must be subtracted because the distance between the farthest limbs is taken.

True Distance (D).

	° ′ ″	″		° ′			
(*d*)	89 35 46	(+ 14)	=	89 36			
(*s*)	30 25 21	(— 21)	=	30 25	sec	10·064308	
(*m*)	39 25 56	(— 56)	=	39 25	sec	10·112074	
				2)159 26			
			x	79 43	cos	9·251677	
			d	89 36			
			(*d* — *x*)	9 53	cos	9·993506	
(S)	30 23 44	(— 21)	=	30 23 23	cos	9·935812	
(M)	40 7 49	(— 56)	=	40 6 53	cos	9·883523	
				2)70 30 16		2)19·240900	
			½ (M + S)	35 15 8	sin	9·620450	
			θ	24 39 51			
			[½ (M + S) + θ]	59 54 59	cos	9·700066	
			[½ (M + S) — θ]	10 35 17	cos	9·992542	
						2)19·692608	
				° ′ ″			
			½ D	44 35 0	sin	9·846304	
				2			
				89 10 0			
			Seconds rejected	— 14			
			True dist D =	89 9 46			

True Greenwich Date.

D =	89° 9′ 46″		Prop¹ log
Dist at VI. =	90 22 10	—	3584
Diff	1 12 24	—	3955
	2ʰ 45ᵐ 15ˢ	=	371
add	6		
	9ᵈ 8 45 15		
Corⁿ for second diffˢ	+ 4		
	9 8 45 19		Greenwich Date.

Longitude.

Star's RA.	Star's Dec.	Mean Sun's RA.	RA of Mer.
19ʰ 44ᵐ 44ˢ·7	N 8° 32′ 15″	1ʰ 12ᵐ 17ˢ·1	M. T. P 14ʰ 30ᵐ
	90	1 18·8	RA of M Sun 1 14
	p 81 27 45	7·4	RA of Mer 15 44
		1 13 43·3	

	° ′ ″		
Altitude (a)	30 23 44		
Latitude (l)	27 30 0	sec	10·052071
Polar dist (p)	81 27 45	cosec	10·004840
	2)139 21 29		
S	69 40 44	cos	9·540670
a	30 23 44		
S−a	39 17 0	sin	9·801511
			2)19·399092
$\frac{1}{2}h$	30° 2′ 38″	sin	9·699546
	2		
	60 5 16		
	4		
	6,0)24,0 21 4		
Easterly Hour ∠ because RA of star is greater than RA of mer	4 0 21·1		
	24 0 0		
Westerly Hour ∠	19 59 38·9		
Star's RA	19 44 44·7		
	39 44 23·6		
RA of mean sun	1 13 43·3		
True mean time at place	14 30 40·3		
True Greenwich Date	8 45 19·0		
	5 45 21·3		
	60		
	4)345 21 18		
Longitude	86 20 19·5 East		

Lunar Observations (Planet).

October 30th, 1876, at about 8^h 20^m P.M., mean time, in lat 51° 20′ N, longitude by account 2° 30′ W, the following lunar was taken to determine the longitude.

Obs alt Saturn, (LL).	Obs alt ☾	Obs dist. NL.
25° 5′ 0″	40° 12′ 0″	39° 15′ 0″
IE + 1 30	IE + 1 6	IE + 1 51

Height of the eye 18 feet.

Greenwich Date nearly.	Longitude.
d h m s	° ′
30 8 20 0	2 30 W
10 0	4
30 8 30 0	10 0

☾ SD.	Change.	☾ HP.	Change.
′ ″	″	′ ″	″
15 34·5	+6·3	57 3·9	22·8
+ 4·4	8·5	+ 16·1	8·5
15 38·9	315	57 20·0	1140
Aug + 9·7	504	Redⁿ 7·2	1824
15 48·6	12)53·55	57 12·8	12)193·80
	4·4	60	16·1
		HP 3433″	

Altitudes.

	° ′ ″		° ′ ″
Obs alt star	25 5 0	Obs alt ☾	40 12 0
IE	+ 1 30	IE	+ 1 6
	25 6 30		40 13 6
Dip	— 4 11	Dip	— 4 11
	25 2 19		40 8 55
Planet's SD	+ 8	SD	+ 15 49
(*s*)	25 2 27	(*m*)	40 24 44
Ref	— 2 2	Ref	— 1 7
	25 0 25		40 23 37
Planet's Parˣ	+ 1	Parˣ	+ 43 35
(S)	25 0 26	(M)	41 7 12

N.B.—If the Hor[l] Parˣ be large, the Parˣ in alt must be worked as in moon.

☾ Parallax in alt.

Log HP 3433″	= 3·535800
Log cos alt	= 9·881799
6,0)261,5″	= 3·417599
+ 43′ 35″ (*d*)	

	° ′ ″
Obs dist NL	39 15 0
IE	+ 1 51
	39 16 51
Planet's SD	+ 8
☾ SD	+ 15 49
(*d*)	39 32 48

True Distance (D).

	° ′ ″	″		° ′		
(*d*)	39 32 48	(— 48)	=	39 32		
(*s*)	25 2 27	(— 27)	=	25 2	sec	10·042842
(*m*)	40 24 44	(— 44)	=	40 24	sec	10·118308
				2)104 58		
			x	52 29	cos	9·784612
			d	39 32		
			(*x* — *d*)	12 57	cos	9·988811

(S) 25 0 26 (— 27) =	24 59 59	cos	9·957277
(M) 41 7 12 (— 44) =	41 6 28	cos	9·877069
	2)66 6 27		2)19·768919
½ (M + S)	33 3 13	sin	9·884459
θ ..	50 1 57		
[½ (M + S) + θ] ..	83 5 10	cos	9·080546
[θ — ½ (M + S)]	16 58 44	cos	9·980645
			2)19·061191
½ D =	19° 50′ 6″	sin	9·530595
	2		
	39 40 12		
Seconds rejected	+ 48		
D =	39 41 0		

True Greenwich Date.

			Prop[1] log.
D =	39° 41′ 0″		
Dist at VI =	38 12 7	—	2556
Diff	1 28 53	—	3064
	2h 40m 7s	=	508
Add	6		
30d	8 40 7		
Cor for second diffs	— 4		
True Greenwich Date	30 8 40 3		

Longitude.

Planet's Ra.	Change.	Planet's Dec.	Change.	Mean Sun's Ra.
h m s 22 16 21·10	s — ·10	° ′ ″ S 12 40 52	″ — ·2	h m s 14 36 34·6
·02	·2	90	·2	1 18·8
22 16 21·08	·020	102 40 52	·04	6·6
				14 38 0·0

N.B. The changes in *Ra* and dec of planet are multiplied by the hours of the longitude in time, but the change here is small.

		° ′ ″		
	Altitude (a)	25 0 26		
	Latitude (l)	51 20 0	sec	10·204267
	Polar dist (p)	102 40 52	cosec	10·010724
	2)	179 1 18		
Ra of Meridian	S	89 30 39	cos	7·931272
——	a	25 0 26		
MTP 8^h 20^m	(S−a)	64 30 13	sin	9·955501
M Sun's Ra 14 38		° ′ ″		2)18·101764
22 58	($\frac{1}{2}h$)	6 27 19	sin	9·050882
		2		
Star west of mer^n because its Ra is less than Ra of mer^n		12 54 38		
		4		
	6,0)	5,1 38 32		
Westerly Hour ∠		0 51 38·5		
Star's Ra		22 16 21·1		
Ra of meridian		23 7 59·6		
Mean Sun's Ra		14 38 0·0		
True mean time at place		8 29 59·6		
True Greenwich Date		8 40 3·0		
		0 10 3·4		
		60		
	4)	10 3 24		
Longitude		2° 30′ 51″	West	

EXAMPLES FOR EXERCISE.

1. January 3rd, 1876, at about 3^h 40^m P.M. mean time nearly, in lat 30° 28′ N. long by account, 72° 10′ W, the following lunar was taken to determine the longitude.

Obs alt ☉	Obs alt ☾	Obs dist
15° 32′ 40″	51° 40′ 20″	79° 54′ 0″
IE − 1 18	IE + 1 10	IE − 1 12

Height of the eye 13 feet.

Ans. Longitude 72° 16′ 30″ W.

2. February 16th, 1876, at about 8^h 30^m A.M., mean time nearly, in lat 30° 47′ S, long by account, 175° 40′ E, the following lunar was taken, to determine the longitude.

Obs alt ☉	Obs alt ☾	Obs dist
35° 20′ 20″	38° 30′ 30″	104° 57′ 30″
IE + 1 10	IE + 1 11	IE − 1 14

Height of the eye 16 feet.

Ans. Longitude 175° 45′ 0″ E.

3. January 8th, 1876, at about $10^h\ 30^m$ P.M., mean time nearly, in lat 29° 30′ N, long by account, 10° 20′ E, the following lunar was taken, to determine the longitude.

Obs alt Regulus	Obs alt ☾	Obs dist FL
27° 50′ 0″	78° 17′ 0″	73° 51′ 0″
IE + 1 51	IE + 1 28	IE + 1 54

Height of the eye 19 feet.

Ans. Longitude 10° 30′ 0″ E.

4. June 6th, 1876, at about $10^h\ 20^m$ P.M., mean time nearly, in lat 21° 18′ S, long by account, 78° 30′ E, the following lunar was taken, to determine the longitude.

Obs alt Spica	Obs alt ☾	Obs dist NL
58° 31′ 0″	70° 17′ 0″	50° 40′ 0″
IE + 1 10	IE + 1 10	IE + 1 35

Height of the eye 13 feet.

Ans. Longitude 78° 40′ 0″ E.

5. January 7th, 1876, at about $6^h\ 10^m$ P.M., mean time, in lat 39° 30′ N, long by account 81° 20′ W, the following lunar was taken to determine the longitude.

Obs alt Mars LL	Obs alt ☾	Obs dist NL
40° 30′ 0″	55° 33′ 0″	67° 28′ 0″
IE + 1 4	IE + 1 25	IE + 1 0

Height of the eye 20 feet.

Ans. Longitude 81° 30′ 0″ W.

Elements for preceding Examples from Nautical Almanac, 1876.

1. ☾'s SD on 3rd noon 15′ 34″ Change + 6″·1 : Hor par 57′ 1″·8, Ch + 22·4″.
   Sun's semi diam 16′ 18″. Sun's dec S 22° 52′ 4″, Ch — 14″·34.
   Eqn time $4^m\ 34^s{\cdot}8$ Ch + $1^s{\cdot}159$ + to app time.
   Dist at VI 79° 3′ 49″ ; Prop log 2963.

2. ☾'s SD on 15th noon 15′ 5″·8, Ch — 4″·7 : Hor par 55′ 18″·6, Ch — 17″·4.
   Sun's Semi diam 16′ 13″. Sun's dec S 12° 47′ 38″, Ch — 51″·35.
   Eqn time $14^m\ 23^s{\cdot}8$ Ch — $\cdot120^s$, + to app time.
   Dist at VI 106° 5′ 28″ ; Prop log 3308.

3. ☾'s SD on 8th noon 16′ 34″·2, Ch + 3″·4 : Hor par 60′ 42″·6, Ch + 12″·3.
   Star's RA $10^h\ 1^m\ 47^s{\cdot}3$, dec 12° 34′ 19″ N, sid time $19^h\ 9^m\ 34^s{\cdot}0$
   Dist at IX 73° 56′ 6″; Prop log 2043.

4. ☾'s SD on 6th noon 14′ 48″·3 Ch — 1″·6 : Hor par 54′ 14″·4, Ch — 5″·8
   Star's RA $13^h\ 18^m\ 41^s{\cdot}7$, dec 10° 31′ 5″ S, sid time $5^h\ 0^m\ 57^s{\cdot}4$.
   Dist at III 49° 36′ 48″; Prop log 3030.

5. (* SD on 7th noon 16′ 24″·8, Ch + 5″·1 : Hor par 60′ 8″·1 Ch + 18″·7. Star's RA 23ʰ 36ᵐ 46ˢ·5, Ch + 6ˢ·5, dec 3° 3′ 11″ S, Ch + 45″·5. Semi diam 4″ Hor par 5″·9 : Sidereal time 19ʰ 5ᵐ 37ˢ·5. Dist at IX 65° 48′ 14″, Prop log 2334.

(13) To compute the Altitude of a Celestial Object.

Where the latitude of a place, the object's polar distance, and hour angle are known, its altitude may be computed, as follows.—

$$\text{If } \sin^2\theta = \sin pd \,.\, \sin \text{co lat} \,.\, \cos^2\tfrac{1}{2} \text{ Hour Angle.}$$

$$\text{Then } \sin^2\frac{\text{Zen dist}}{2} = \sin\left\{\frac{pd + \text{colat}}{2} + \theta\right\}\sin\left\{\frac{pd + \text{colat}}{2} - \theta\right\}$$

The Hour Angle is determined as follows.—

1) If the object observed be the **Sun**, its hour angle is the **true apparent time** at the place of observation.

(2) If a **star**, or the **moon** be observed, then Hour ∠ = Mean Time at place + Mean Sun's Ra — Star's Ra. With these two rules, no difficulty need be experienced.

EXAMPLE I.

Compute the apparent alt of the Sun's centre on January 18th, 1876, in latitude 42° 10′ N, longitude 73° 10′ E, at 10ʰ 45ᵐ 30ˢ A.M., mean time at place.

Note. (*a*). Find the Greenwich Date.

(*b*) Correct the declination of the object, and if the time given be mean time, the Equation of Time also.

(*c*) Hour Angle = MTP ± Equation of Time or **True appt** Time at place.

(*d*) Colatitude, found by subtracting the latitude from 90°.

(*e*) Application of formula.

Greenwich Date.	Longitude.	Sun's Declination.	Change.
d h m s	° ′	° ′ ″	″
17 22 45 30	73 10 E	S 20 48 58	— 29·31
— 4 52 40	4	— 8 44	17· 9
17 17 52 50	6,0)29,2 40	20 40 14	26379
	4 52·40	90 0 0	20517
		p 110 40 14	2931
			6,0)52,4,649
			— 8 44·6

Equation of Time.	Change.	Hour Angle.			
m s			h	m	s
10 13·98	+·832	Mean time place	22	45	30
+ 14·89	17·9	Equation of time	—	10	28·9
10 28·87	7488		22	35	1·1
— to MT	5824	Subtract from	24	0	0
	832		1	24	58·9
	14,8928		60		
			4)84	58	54
			2)21	14	43
		½h	10	37	21

Colatitude.

Lat	42°	10′ N
	90	0
Colat = l′	47	50

Computation.

$$\sin^2 \theta = \sin p . \sin l' \cos^2 \tfrac{1}{2}h$$

$$\sin^2 \frac{z}{2} = \sin \left\{ \frac{p + l'}{2} + \theta \right\} \sin \left\{ \frac{p + l'}{2} - \theta \right\}$$

	°	′	″		
½h =	10	37	21	cos	9·992493
					2
					19·984986
p =	110	40	14	sin	9·971102
l′ =	47	50	0	sin	9·869933
	2)158	30	14		2)19·826021
½(p + l′)	79	15	7	sin	9·913010
θ	54	56	0		
½(p + l′) + θ	134	11	7	sin	9·855574
½(p + l′) — θ	24	19	7	sin	9·614697
					2)19·470271
½z	32	55	0	sin	9·735135
			2		
True Zen dist	65	50	0		
	90	0	0		
True alt ⊙′ cr	24	10	0		
Parallax	—		8		
	24	9	52		
Ref^{n}	+	2	7		
Apparent alt ⊙′ cr	24	11	59		

EXAMPLE II.

Compute the altitude of Aldebaran, on January, 8th 1876, at $8^h\ 42^m\ 30^s$ P.M., mean time, in latitude 39° 20′ N, longitude 84° 30′ W.

Note.— (*a*) Find the Greenwich Date.

(*b*) Correct the Mean Sun's *Ra*, also take out the *Ra* and dec of star: find polar distance (*p*).

(*c*) Hour Angle = MTP + Mean Sun's *Ra* — Star's *Ra*, reduced to degrees.

(*d*) Colatitude.

(*e*) Computation as before.

Greenwich Date.	Longitude.	Mean Sun's *Ra*.	Star's *Ra* and Dec
d h m s	° ′	h m s	
8 8 42 30	84 30 W	19 9 34·02	*Ra* $4^h\ 28^m\ 49^s·5$
+ 5 38 0	4	2 17·99	Dec 16° 15′ 38″ N
8 14 20 30	6,0)33,8 0	8·28	90
	5 38·0	·08	*p* 73 44 22
		19 11 55·37	

Hour Angle.

	h m s
Mean time place	8 42 30
Mean Sun's *Ra* +	19 11 55·4
	27 54 25·4
Star's *Ra*	4 28 49·5
	23 25 35·9
Subtract from	24 0 0
2)	0 34 24·1
	0 17 12·0
	60
4)	17 12 0
½ *h*	4 18 0

Colatitude.

	° ′
Lat =	39 20 N
	90 0
l′ =	50 40

Computation.

$$\sin^2 \theta = \sin p . \sin l' . \cos^2 \tfrac{1}{2} h$$

$$\sin^2 \frac{z}{2} = \sin \left\{ \frac{p + l'}{2} + \theta \right\} \sin \left\{ \frac{p + l'}{2} - \theta \right\}$$

	° ′ ″		
$\frac{1}{2} h$	4 18 0	cos	9·998776
			2
			19·997552
p	73 44 22	sin	9·982270
l'	50 40 0	sin	9·888444
	2)124 24 22		2)19·868266
$\frac{1}{2}(p + l')$	62 12 11	sin	9·934133
θ	59 14 9		
$[\frac{1}{2}(p + l') + \theta]$	121 26 20	sin	9·931049
$[\frac{1}{2}(p + l) - \theta]$	2 58 2	sin	8·714033
			2)18·645082
$\frac{1}{2} z$	12 7 58	sin	9·322541
	2		
True zenith distance	24 15 46		
	90 0 0		
True alt	65 44 14		
Ref	+ 26		
App alt	65 44 40		

EXAMPLE III.

January 9th, 1876, at 8ʰ 40ᵐ 30ˢ P.M., mean time, in lat 47° 20′ N, long 67° 30′ W, compute the true and apparent altitudes of the moon's centre.

Note.—(*a*) Find the Greenwich date.

(*b*) Correct for this date the moon's S.D. and Horizontal Parallax, also her R*a*, dec, and the mean sun's R*a*.

(*c*) Hour ∠ = mean time place + mean sun's R*a* — moon's R*a*.

(*d*) Colatitude.

(*e*) Computation as before.

Greenwich Date.	Longitude.	Mean Sun's R*a*.
d h m s	° ′	h m s
9 8 40 30	67 30 W	19 13 30·6
+ 4 30 0	4	2 8·1
9 13 10 30	6,0)27,0 0	1·6
	4 30 0	·1
		19 15 40·4

Moon's Ra.	Change.	Moon's Declination.	Change.
h m s	"	° ' "	"
6 8 24·7	+2·87	N 28 27 42	—·54
+ 30·1	10·5	— 6	10·5
6 8 54·8	1435	28 27 36	270
	287·	90 0 0	54·
	30·135	*p* 61 32 24	5·670

Moon's S.D.	Change.	Moon's Hor' Par*	Change.
' "	"	' "	"
16 40·9	—·2	61 7·1	—·8
*Aug + 14·8	1·2	·1	1·2
16 55·7	12)·24	61 7·0	12)·96
	·02	Red — 6·6	·08
		61 0·4	
		60	
		HP 3660	

Hour Angle.

	h m s
Mean time place	8 40 30
Mean Sun's Ra +	19 15 40·4
	27 56 10·4
Moon's Ra —	6 8 54·8
	21 47 15·6
	24 0 0
2)	2 12 44·4
	1 6 22·2
	60
4)	66 22 12
½*h*	16 35 33

Colatitude.

Lat	47° 20'
	90 0
l' =	42 40

* The Augmentation is filled in when the Moon's altitude is found.

Computation.

$\frac{1}{2}h$	16° 35′ 33″	cos	9·981528
			2
			19·963056
p	61 32 24	sin	9·944063
l'	42 40 0	sin	9·831058
	2)104 12 24		2)19·738177
$\frac{1}{2}(p + l')$	52 6 12	sin	9·869088
θ	47 42 38		
$[\frac{1}{2}(p + l') + \theta]$	99 48 50	sin	9·993598
$[\frac{1}{2}(p + l') - \theta]$	4 23 34	sin	8·884190
	° ′ ″		2)18·877788
$\frac{1}{2}z$	15 56 44	sin	9·438894
	2		
True zenith distance	31 53 28		
	90		
True altitude centre	58 6 32		
True alt Moon's cr	58° 6′ 32″		

(1) Approximate Parallax in alt.

Log horizontal parallax 3660 =	3·563481
Cos true alt 58° 6′ =	9·722994
60)193″,4 =	3·286475
Approximate Parallax in alt	32′ 14″

True alt Moon's cr	58° 6′ 32″
Approximate parallax in alt	— 32 14
Approximate app' alt	57 34 18

(2) Correct Parallax in alt.

Log horizontal parallax 3660 =	3·563481
Cos app alt 57° 34′ =	9·729422
60)196″,3 =	3·292903
True Parallax in alt	32′ 43″

True alt Moon's cr	58° 6′ 32″
Parallax in alt	— 32 43
	57 33 49
Ref	+ 37
App' alt Moon's cr	57 34 26

N.B.—If the semi diam be applied, we obtain the apparent altitude of the moon's upper or lower limb.

It is evident that the cosine of the true altitude is not the correct quantity to be applied to the horizontal parallax, thus an approximate parallax is found, and from this, the apparent altitude nearly is obtained, of which the cosine is taken.

(14) Computation of the Longitude by a Lunar Observation when the Altitudes are not observed.

Example.

March 29th, 1876, at 3ʰ 17ᵐ 30ˢ P.M., true mean time at place, in lat 47° 20′ N, long by account 100° 20′ W, the apparent distance between the sun and moon was 52° 48′ 40″, IE + 1′ 11″: find the longitude.

(1) To compute the Sun's Altitude.

Greenwich Date.	Longitude.	Declination.	Change.
d h m s	° ′	° ′ ″	″
29 3 17 30	100 20 W	N 3 37 47	+ 58·31
+ 6 41 20	4	+ 9 43	10
29 9 58 50	6,0)40,1 20	3 47 30	6,0)58,3·10
	6 41·20	90 0 0	9·43
		p 86 12 30	

Equation of Time.	Change.	Hour Angle.	
m s	s		h m s
4 45·31	— ·763	M.T.P.	3 17 30
— 7·63	10	Eq Time	— 4 37·7
4 37·68	7·630		2)3 12 52·3
— to M.T.			1 36 26·1
			60
			4)96 26 6
			½h 24 6 31

Computation.

$\frac{1}{2}h$	24° 6′ 31″	cos	9·960363
			2
			19·920726
p	86 12 30	sin	9·999048
l'	42 40 0	sin	9·831058
	2)128 52 30		2)19·750832
$\frac{1}{2}(p + l')$	64 26 15	sin	9·875416
θ	48 38 37		
$[\frac{1}{2}(p + l') + \theta]$	113 4 52	sin	9·963764
$[\frac{1}{2}(p + l') - \theta]$	15 47 38	sin	9·434851
			2)19·398615
$\frac{1}{2}$ Z	30° 1′ 32″	sin	9·699307
	2		
True zenith distance	60 3 4		
	90 0 0		
True altitude (S)	29 56 56		
Parallax	− 8		
	29 56 48		
Ref	+ 1 38		
(s)	29 58 26		

(2) To Compute the Moon's Altitude.

M Sun's *Ra*.	Moon's *Ra*.	Change.	Moon's Decn.	Change.
0h 28m 55s·0	3h 56m 41s·3	+ 2s·5	N 25° 14′ 46″	+ 8″·7
1 28·7	+ 2 27·0	58·8	+ 8 31	58·8
9·5	3 59 8·3	2940	25 23 17	4116
·1		1176	90 0 0	4704
0 30 33·3		6,0)14,7·00	p 64 36 43	6,0)51,1·56
		2·27		+8·31

	Moon's S D.	Change.	Moon's Hor1 Parx	Change.
	16′ 12″·9	+ ·6″	59′ 24″·4	+ 2″·4
	5	10	+ 2·0	10
	16 13·4	12) 6·0	59 26·4	12)24·0
Aug	+ 14·6	+ ·5	Red − 6·4	+ 2
	16 28·0		59 20·0	
			60	
			HP 3560″	

Hour Angle.

	h	m	s
MTP	3	17	30
Mean Sun's R*a* +	0	30	33·3
	3	48	3·3
Moon's R*a* —	3	59	8·3
	0	11	5·0
	60		
4)	11	5	0
2)	2	46	15
$\frac{1}{2}h$	1	23	7

Colatitude.

	°	′	
Lat =	47	20	N
	90		
l' =	42	40	

Computation.

	°	′	″		
$\frac{1}{2}h$	1	23	7	cos	9·999873
					2
					19·999746
p	64	36	43	sin	9·955892
l'	42	40	0	sin	9·831058
2)	107	16	43	2)	19·786696
$\frac{1}{2}(p+l')$	53	38	21	sin	9·893348
θ	51	28	5		
$[\frac{1}{2}(p+l')+\theta]$	105	6	26	sin	9·984725
$[\frac{1}{2}(p+l')-\theta]$	2	10	16	sin	8·578453
				2)	18·563178
$\frac{1}{2}h$	11	1	32	sin	9·281589
			2		
True zenith distance	22	3	4		
	90	0	0		
True altitude (M)	67	56	56		

Approximate Parallax in alt.

Log HP	3·551450
Cos True alt	9·574512
6,0)133,6 =	3·125962
22′ 16″	

Approximate apparent alt.

True alt (M)	67° 56′ 56″
Parallax nearly	— 22 16
Approx^te^ app alt	67 34 40

True Parallax.

Log HP =	3·551450
Cos app alt =	9·581618
6,0)135,8″ =	3·133068
— 22′ 38′	

Apparent Altitude.

	° ′ ″
True alt (M)	67 56 56
Parallax in alt	— 22 38
	67 34 18
Ref^n^	+ 24
(m)	67 34 42

Apparent Distance (d).

Obs dist	52° 48′ 40″
IE	+ 1 11
	52 49 51
Sun's SD	+ 16 3
Moon's SD	+ 16 28
(d)	53 22 22

True Distance and Longitude.

(d)	53° 22′ 22″	(− 22″) =	53° 22′		
(s)	29 58 26	(− 26) =	29 58	sec	10·062324
(m)	67 34 42	(− 42) =	67 34	sec	10·418382
			2)150 54		
		x	75 27	cos	9·400062
		d	53 22		
		$(x-d)$	22 5	cos	9·966910
(S)	29 56 56	(− 26″) =	29 56 30	cos	9·937786
(M)	67 56 56	(− 42) =	67 56 14	cos	9·574752
			2)97 52 44		2)19·360216
		$\frac{1}{2}(M+S)$	48 56 22	sin	9·680108
		θ	28 36 11		
		$[\frac{1}{2}(M+S)+\theta]$	77 32 33	cos	9·333881
		$[\frac{1}{2}(M+S)-\theta]$	20 20 11	cos	9·972049
					2)19·305930
		$\frac{1}{2}$D	26° 43′ 38″	sin	9·652965
			2		
			53 27 16		
		Seconds rejected	+ 22		
		D	53 27 38		Prop[l] log.
		Distance at IX	52 54 37	—	2564
		Diff	0 33 1	—	7365
			$0^h\ 59^m\ 35^s$	=	4801
		Add	9		
		True Greenwich Date	9 59 35		
		True Mean Time	3 17 30		
			6 42 5		
			60		
			4)402 5 0		
		Longitude	100 31 15	W	

EXAMPLES FOR EXERCISE.

1. November 10th, 1876. at $9^h\ 30^m\ 30^s$ A.M. true mean time at place, in lat 27° 10′ N, long by account 47° 20′ E, the obs dist between the sun and moon was 69° 53′ 0″, IE + 1′ 7″ : Height of the eye 19 feet: required the longitude.

Ans. 47° 31′ 45″.

2. May 6th, at $10^h\ 30^m\ 30^s$ P.M., true mean time at place, in lat 31° 27′ S, long by account 111° 20′ W, the obs dist between Antares and the Moon's farthest limb was 34° 32′ 0″, IE + 1′ 28″ : Height of the eye 20 feet: required the longitude.

Ans. 111° 29′ 45″ W.

Elements for preceding Examples from Nautical Almanac, 1876.

1. Sun's declination S 17° 3′ 28″: change + 42″·49 : Equation of time $16^m\ 0^s$·36 change — ·241 (+mean time): Sun's semi diam 16′ 12″; Moon's RA $10^h\ 39^m\ 37^s$.5, change + 2^s·1 ; Moon's dec 9° 27′ 32″ N ; change — 15″.6 ; Sidereal Time $15^h\ 16^m\ 0^s$·13; Moon's SD 15′ 58″·1, change — 3″·5 : Moon's HP 58′ 30″·4, change — 13″·1.

2, Star's RA $16^h\ 31^m\ 51^s$: Star's dec S 26° 9′ 31″ : Sid time $2^h\ 58^m\ 44^s$·1 ; Moon's RA $13^h\ 59^m\ 46^s$·6 change + 1^s·96; Moon's dec S 15° 21′ 29″, change + 12″·9 ; Moon's SD 15′ 9″·5 change — 3″·5 : Moon's HP 55′ 32″·3, change — 13″·0.

TIME.

(13) To find the Rate of a Chronometer.

This is determined by finding the error on a given day and on a subsequent day. If the difference of these errors be taken, and divided by the time elapsed, the result must be the rate of the chronometer, thus,—

If on August 10th at noon, my chronometer be slow $2^h\ 10^m\ 30^s$·5 for mean time at place and on September 30th, its error be $2^h\ 14^m\ 45^s$ slow, find its daily rate.

Here,

	h m s	
Error on August 10th	2 10 30·5	
Error on September 30th	2 14 45·0	s
Difference	4 14·5	= 255·5

The interval of time is 51 days; then,

$\frac{255^s\cdot5}{51} = 5^s\cdot1$ nearly, and the chronometer is evidently losing, since it is slower on September 30th than on August 10th.

Example 1.—A chronometer is slow for mean time $2^h\ 40^m\ 30^s{\cdot}5$ at place in longitude 62° E: find its error on Greenwich mean time.

	h	m	s
Suppose it to show	12	0	0
Adding the error	2	40	30·5
The chronometer ought to show	14	40	30·5
Subtract the longitude in time	4	8	0
The corresponding Greenwich time is	10	32	30·5
But the chronometer showed	12	0	0
Chronometer fast on G.M.T.	1	27	29·5

Example 2.—A chronometer is slow $1^h\ 40^m\ 30^s{\cdot}5$ at a place A in longitude 79° 30′ W, what is its error for a place B in longitude 100° 30′ W?

	h	m	s
Suppose the chronometer to show	12	0	0
Error for A, slow	1	40	30·5
Time at A	13	40	30·5
Long of A West +	5	18	0
Greenwich Time	18	58	30·5
Longitude of B West −	6	42	0
Time at B	12	16	30·5
Chronometer showed	12	0	0
Error slow for B	0	16	30·5

Example 3.—A chronometer is slow for mean time at A, on July 10th, $15^m\ 47^s.5$ and on December 12th, $3^h\ 17^m\ 42^s$ fast for mean time at B: what was the rate of the chronometer, having given longitude A 21° 30′ W. and longitude B 62° 10′ W.

	h	m	s
1. Suppose the chronometer to show	12	0	0
Slow at A on July 10th		15	47·5
Mean time at A	12	15	47·5
Longitude A (west)	1	26	0
Greenwich Time	13	41	47·5
Chronometer showed	12	0	0
Slow on G.M.T., July 10th	1	41	47·5

	h m s
2. Suppose the chronometer to show	12 0 0
Fast at B August 12th	3 17 42
Mean Time at B	8 42 18
Longitude B (west)	4 8 40
Greenwich time	12 50 58
Chronometer showed	12 0 0
Slow on G.M.T., December 12th	50 58

	h m s
Therefore slow on G.M.T., July 10th	1 41 47·5
„ „ Dec. 12th	0 50 58·0
	0 50 49·5
	60
Gained in 155 days	3049·"5

or 19·"7 per day.

(14) To find the Error of Chronometer by Equal Altitudes of the Sun.

The equation of equal altitudes is a correction to be applied to the middle of the two times shown by a chronometer when the sun had equal altitudes, in order to obtain the time the chronometer showed at the instant that the sun's centre was on the meridian of the place of observation.

As the sun's declination is continually changing, it follows that the hour angles obtained from two equal altitudes on the same day, are unequal; the easterly or westerly hour angle as the case may be, is the larger, and half the difference of these hour angles is actually the equation of equal altitudes.

Example.

On July 12th, 1876, in latitude 42° 20′ S, long 37° 20′ W, when a chronometer showed 9^h 47^m 10^s, and 3^h 10^m 30^s, the sun had equal altitudes: find the error of chronometer.

Note.—(*a*) Find the middle time by chronometer, also the half elapsed time, which reduce to arc.

(*b*) Find the Greenwich date of apparent noon, for which, correct the sun's declination and the equation of time.

(*c*) $C = \dot{\text{change}}$ in declination × half elapsed time.

(*d*) Compute the equation of equal altitudes and apply to the middle time; the difference between this and apparent noon is the error for apparent time at place.

Middle Time.	Half Elapsed Time.	Greenwich Date.	Longitude.
h m s	h m s	d h m s	° ′
21 47 10	21 47 10	12 0 0 0	37 20 W
27 10 30	27 10 30	+ 2 29 20	4
2)48 57 40	2)5 23 20	12 2 29 20	6,0)14,9 20
24 28 50	2 41 40		2 29 20
	60		
	4)161 40		
	H 40 25		

Declination.	Change.	Equation of Time.	Change.
° ′ ″	″	m s	s
N 21 54 26	— 21·39	5 21·91	+ ·305
— 53	2·5	·76	2·5
21 53 33	10695	5 22·67	1525
	4278	+ App T	610
	53·475		·7625

C = change in dec (21″·39) multiplied by the half elapsed time $2^h·7$, and

$$21·39 \times 2·7 = 57·753$$

A = *c* tan lat cosec H.		B = *c* tan dec cot H.	
Log *c*	1·761552	Log *c*	1·761552
Tan lat	9·959516	Tan dec	9·603858
Cosec H	10·188196	Cot H	10·069780
A 81·15 =	1·909264	B 27·23 =	1·435190

When the latitude and declination are opposite in name, these are added: if same name, subtracted. Here latitude is S, and declination N, therefore

A	81·15
B	27·23
	15)108·38
Equation of equal altitudes	7ˢ ·2

This must be applied to the middle time as follows.

Fig. 25.

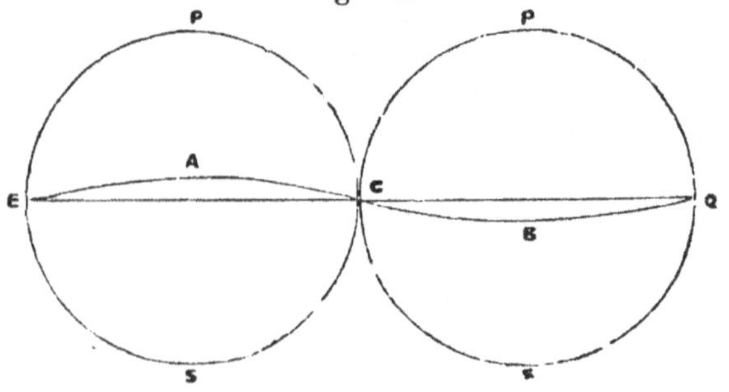

K

Let the two circles represent the hemispheres of the heavens, EQ the equator projected as a straight line: then EABQ will be the Ecliptic, or Sun's imaginary path in the heavens during the year. Now on March 21st, he will be at E the first point of Aries: on June 21st, at A: September 21st, at C: and December 21st, at B.

In this case, the month is July, the latitude is South, the sun is therefore travelling towards the South pole: hence his south polar distance is decreasing, therefore the correction computed is subtractive from the Middle Time.

	h	m	s	
(1) Middle time by chron was	24	28	50	
Equation of equal altitudes −			7·2	
Chron showed at app noon	24	28	42·8	
But it ought to have shown	24	0	0	
Therefore its error on app time place		28	42·8	fast.

	h	m	s	
(2) Apparent noon	24	0	0	
Equation of time	+	5	22·7	
Chron would show at noon if right for MTP	24	5	22·7	
But it shows	24	28	42·8	
Therefore at mean noon it was		23	20·1	fast.

	h	m	s	
(3) Again chron if right for MTP would show	24	5	22·7	
Longitude +	2	29	20	
Corresponding Greenwich Time	26	34	42·7	
Chron shows	24	28	42·8	
Error on GMT	2	5	59·9	slow.

EXAMPLES FOR EXERCISE.

1. January 10th, 1876, in latitude 54° 18′ S, longitude 120° W, when a chronometer showed 8^h 31^m 10^s, and 2^h 45^m 30^s, the sun had equal altitudes: find the error of chronometer on Mean and Apparent Time at place.

Ans. On Mean Time 29^m $19^s·6$ slow.
On App^t Time 21 33 ·1 slow.

2. February 7th, in latitude 34° 18′ N, longitude 72° 30′ E, when a chronometer showed 9^h 31^m 17^s, and 2^h 31^m 45^s, the sun had equal altitudes: find the error of chronometer on Mean and Apparent Time at place.

Ans. On Mean Time 13^m $2^s·8$ slow.
On App^t Time 1 19 ·7 fast.

3. March 4th, in latitude 57° 30′ N, longitude 69° 10′ W, when a chronometer showed 10^h 31^m 10^s, and 3^h 15^m 40^s, the sun had equal altitudes: find the error of chronometer on Mean and Apparent Time at place.

Ans. On Mean Time 41^m 13^s fast.
On App^t Time 52 58·5 ,,

Elements for preceding Examples from Nautical Almanac, 1876.

1. Sun's dec S 22° 0′ 58″, change — 22″·08; Equation of time $7^m\ 38^s{\cdot}4$, change + $1^s.016$, + to app time.

2. Sun's dec S 15° 43′ 7″, change — 46″·02; Equation of time $14^m\ 19^s{\cdot}3$, change + .169, + to app time.

3. Sun's dec S 6° 10′ 51″, change — 57″·86; Equation of time $11^m\ 48^s{\cdot}4$, change — ·604, + to app time.

(15) To find the Error of a Chronometer by Equal Altitudes of a Fixed Star.

In this case as the declination of a star changes so slowly there is no Equation of Equal Altitudes. The chronometer time of the star's transit over the meridian of the place, is the mean or middle of the two chronometer times shown when the star had equal altitudes. If, by an independent method, we can find the actual time of the star's passing the meridian, then the difference between this and the middle time is the error.

Example.

On January 12th 1876, in longitude 80° 30′ W, when a chronometer showed $7^h\ 31^m\ 10^s$, and $9^h\ 24^m\ 20^s$, the star Aldebaran had equal altitudes: find the error.

Note.—(*a*) Find the middle time by chronometer.

(*b*) Compute the approximate time of the star's transit, which correct for longitude.

(*c*) Correct the sidereal time for this date.

(*d*) Then from the star's (+ 24^h if necessary), subtract the mean sun's RA, the result is the correct time of transit.

(*e*) The difference between this and the middle time is the error.

Chronometer Time of Star's Transit.

	h	m	s
	7	31	10
	9	24	20
2)	16	55	30
Star was on mer by chron	8	27	45

To find the Approximate Time of Star's Transit.

	d	h	m	s
From Star's Ra		4	28	49·5
Subtract sidereal time		19	25	20·2
Date at place, January	12	9	3	29·3
Longitude +		5	22	0
Greenwich date	12	14	25	29·3

Mean Sun's R*a*.

		h	m	s
Sidereal time		19	25	20·2
Acceleration for	14^h =		2	17·9
	25^m =			4·1
	29^s =			·1
		19	27	42·3

To find True Time of Star's Transit.

	h	m	s	
From Star's R*a*	4	28	49·5	
Subtract mean Sun's R*a*	19	27	42·3	
Star passed meridian at	9	1	7·2	
Chronometer time of transit	8	27	45·0	
Error		33	22·2	slow on M.T.P.

EXAMPLES FOR EXERCISE.

1. February 18th, 1876, in longitude 106° 30′ E, when a chronometer showed $6^h\ 40^m\ 40^s$ and $10^h\ 50^m\ 30^s$, Rigel had equal altitudes: find the error.

Ans. $1^h\ 29^m\ 16^s$ fast.

2. March 12th, in longitude 79° 20′ W, when a chronometer showed $5^h\ 26^m\ 10^s$ and $7^h\ 31^m\ 40^s$, Castor had equal altitudes: find the error.

Ans. $1^h\ 33^m\ 42^s{\cdot}5$ slow.

3. April 21st, in long 110° 30′ E, when a chronometer showed $7^h\ 31^m\ 16^s$ and $10^h\ 44^m\ 20^s$, Regulus had equal altitudes: find the error.

Ans. $1^h\ 5^m\ 42^s$ fast.

ELEMENTS FOR PRECEDING EXAMPLES FROM NAUTICAL ALMANAC, 1876.

1. R*a* of Rigel	5^h	7^m	$32^s{\cdot}9$.	Sidereal time	21^h	51^m	$12^s{\cdot}8$.
2. R*a* of Castor	7	26	42·9.	Sidereal time	23	21	53·6.
3. R*a* of Regulus	10	1	47·7·	Sidereal time	1	59	35·8.

(16) TO FIND THE ERROR OF CHRONOMETER FROM AN ALTITUDE OF THE SUN.

Altitudes taken for the purpose of deducing time at ship, produce the best results when the object observed is far from the meridian, or near to the Prime Vertical Circle, inasmuch as any error in the altitude then produces but a small effect upon the hour angle deduced from it. On the contrary, sights for the latitude should be taken either when the body is on the meridian or near to it, when care and deliberation are necessary in observing the altitudes. This method of finding the error is similar to that already given in finding the longitude.

EXAMPLE.

On September 21st, 1876, on board H.M.S. "Worcester," Greenhithe, (latitude 51° 27′ 10″ N, longitude 15′ E), the following sights were taken to find the error of the chronometer.

	Chronometer.	Sextant $\underline{\odot}$ (art[1] horizon).
	d h m s	° ′ ″
September	21 2 36 39	56 12 10
	37 8	56 6 30
	37 21	56 0 30
	38 0	55 55 0
	38 28	55 47 30

Index Error of Sextant — 1′ 45″.

	Mean of Times.	Mean of Altitudes $\underline{\underline{\odot}}$
	21ᵈ 2ʰ 37ᵐ 30ˢ	56° 0′ 44
Approx longitude	— 1 0	
	12 2 36 30	

Declination.	Change.	Equation of Time.	Change.
N 0° 28′ 14″·5	— 58″·46	7ᵐ — 9ˢ·36	+·870
— 2 32·0	2·6	+ 2·26	2·6
0 25 42·5	35076	7 11·62	5220
90 0 0	11692	— to app T.	1740
p 89 34 17·5	6,0)151·996		2·2620
	2·32		

Corrected Altitude.

	° ′ ″
$\underline{\underline{\odot}}$	56 0 44
IE	— 1 45
	2)55 58 59
	27 59 29
Ref	— 1 47
	27 57 42
S.D.	+ 15 59
	28 13 41
Par[x]	+ 8
(*a*)	28 13 49

Hour Angle.

$$\text{Sin } \tfrac{1}{2}h = \sqrt{\text{Sec } l \text{ cosec } p \,.\, \cos \text{S} \,.\, \sin(\text{S}-a)}$$

	° ′ ″		
a	28 13 49		
l	51 27 10	sec	10.205400
p	89 34 17	cosec	10·000012
	2)169 15 16		
S	84 37 38	cos	8·971439
a	28 13 49		
(S — *a*)	56 23 49	sin	9·920588
			2)19·097439
$\tfrac{1}{2}h$	20 43 5	sin	9·548719

Error of Chronometer.

	° ′ ″
½h	20 43 5
	2
	41 26 10
	4
	6,0)16,5 44 40
True app time place	2 45 44·7
Equation of time	— 7 11·6
True mean time place	2 38 33·1
Chronometer showed	2 37 30·0
Error	1 3·1 slow on M.T.P.

Again,

	h m s
True apparent time	2 45 44·7
Chronometer showed	2 37 30·0
Error	8 14·7 slow on A.T.P.

(17) To find the Error of Chronometer from an Altitude of a Star.

If on September 14th, 1876, in lat 51° 20′ N, long 70° W, the following altitudes of the star α Cygni be observed, find the error of chronometer.

Chronometer. h m s	Sextant. ° ′ ″
7 5 30	53 2 10
5 50	5 30
6 15	7 40
6 35	10 10
6 55	12 50

Index error of sextant — 47″: Height of the eye 24 feet.

Mean of Times.

September	14ᵈ	7ʰ	6ᵐ	13ˢ
Longitude	+	4	40	0
G.D. nearly	14	11	46	13

Mean of Altitudes.

53° 7′ 40

Mean Sun's Ra.

11ʰ 35ᵐ 13ˢ·1
1 48·4
7·5
·1
11 37 9·1

Star's Ra.

20ʰ 37ᵐ 14ˢ·7

Star's declination.

N	49° 50′ 34″
	90 0 0
p	40 9 26

Corrected Altitude.

	° ′ ″
Obs alt	53 7 40
IE	— 47
	53 6 53
Dip	— 4 49
	53 2 4
Ref	— 43
(*a*)	53 1 21

Hour Angle.

$$\text{Sin } \tfrac{1}{2}h = \sqrt{\text{Sec } l \text{ cosec } p \,.\, \text{Cos } s \,.\, \text{Sin } (S-a)}$$

	° ′ ″		
(*a*)	53 1 21		
(*l*)	51 20 0	sec	10·204267
(*p*)	40 9 26	cosec	10·190517
2)	144 30 47		
S	72 15 23	cos	9·483963
a	53 1 19		
(S—*a*)	19 14 4	sin	9·517763
			2)19·396510
½*h*	29 56 44	sin	9·698255

Error of Chronometer.

½*h*	° ′ ″ 29 56 44	
	2	
	59 53 28	
	4	
6,0)	23,9 33 52	
Easterly Hour Angle	3 59 33·4	
	24 0 0	
Westerly Hour Angle	20 0 26·6	
Star's Ra	20 37 14·7	
Ra of Meridian	40 37 41·3	
Mean Sun's Ra	11 37 9·1	
True Mean Time at place	5 0 32·2	
Chronometer showed	7 6 13·0	
Error	2 5 40·8	fast on MTP.

Since the Error is so large it will be now requisite to correct the Mean Sun's Ra again for a new date, therefore.

	d h m s
Greenʰ date nearly	14 11 46 13
Error	— 2 5 40·8
True GD	14 9 40 32·2

Correct Mean Sun's Ra.

h m s
11 35 13·1
1 18·8
6·6
·1
11 36 38·6

	h	m	s	
Ra of Meridian	40	37	41·3	
Correct Mean Sun's Ra	11	36	38·6	
True Mean Time	5	1	2·7	
Chronometer showed	7	6	13·0	
True Error	2	5	10·3	fast on M.T.P.

(18) TO FIND THE ERROR IN AN HOUR ANGLE FROM AN ERROR IN THE OBSERVED ALTITUDE.

EXAMPLE.

In lat 30° N, when the Sun's Azimuth was N 50° E, an error of 30″ was made in the altitude: find the corresponding error in the hour angle.

Rule, Error in hour angle = Error in alt × sec lat × cosec Az

" " = 30″ × sec 30° × cosec 50°

Log 30 = 1·477121
Sec 30 = 10·062469
Cosec 50 = 10·115746
15)45″·2 = 1·655336

Error in hour angle 3·01ˢ

(19) TO FIND THE RATE OF A CHRONOMETER BY LUNARS.

This consists in finding the Greenwich dates from a series of Lunar observations,—from observed distances and altitudes, and by comparing these respective dates with the times shown by the chronometer when the observations were taken, from which the chronometer's error on Greenwich Mean Time is known each day, and thus, the rate of the chronometer.

(20) TO FIND THE ALTITUDES WHEN A LUNAR DISTANCE IS TAKEN, FROM ALTITUDES OBSERVED BEFORE AND AFTER.

If the altitudes be considered as changing uniformly, we have as follows,—

$$\frac{\text{Interval of time between alts}}{\text{Interval of time between 1st alt and dist}} = \frac{\text{Change of alt in 1st int}^{\text{l}}}{\text{Correction of 1st alt}}$$

If these be designated by A, B, C, D, respectively, we have

$$\frac{A}{B} = \frac{C}{D} \text{ or } D = \frac{BC}{A}$$

EXAMPLE.

(1) At $6^h\ 42^m\ 30^s$ by chron the Moon's alt was 51° 2′ 30″
(2) At 6 47 15 " " dist between ☾ and ☉ " 72 14 10
(3) At 6 51 20 " " the Moon's alt " 51 24 30

Find the Mon's alt at $6^h\ 47^m\ 15^s$.

The interval between (1) and (3) is $8^m\ 50^s = 530^s = A$.
" " " (1) " (2) " 4 45 = 285 = B.
Diff of altitudes at (1) " (3) " 22′ 0″ = 1320 = C.

$$\text{Then } D = \frac{BC}{A} = \frac{1320 \times 285}{530} = 11'\ 49''$$

And as the moon's altitude between (1) and (3) is increasing, this correction will be added, therefore

Moon's first alt =	51° 2′ 30″
Correction	+ 11 49
Moon's alt at $6^h\ 47^m\ 15^s$ =	51 14 19

The same may be done for sun or star.

LATITUDE AND LONGITUDE.

(21) Latitude and Longitude by Sumner's Method.

This method, by Captain Sumner of the United States Navy, who published it in 1843, consists in determining the place of the ship by the intersection of two "lines of position."

The approximate latitude being known, assume one differing from it some minutes less, and another the same number of minutes more.

Altitudes of the sun are taken, some considerable interval being allowed to elapse, as in Ivory's method: the work then proceeds as follows:—

1. With the first altitude, first assumed latitude, and the corrected polar distance from the first Greenwich date of the observation, work an assumed longitude which call (a).
2. Again with the first altitude, second assumed latitude, and first polar distance deduce another, (a').
3. With the second altitude, first latitude, and polar distance deduced from the second Greenwich date, calculate (b).
4. With the second altitude, second latitude, and second polar distance, deduce (b').

The lines joining $a\ a'$, and $b\ b'$ on a chart, are "lines of position," or more properly, on the earth's surface, "circles of equal altitudes." The intersection of these lines gives the true position of the ship.

If the ship changes her place in the interval between the observations, the correction for ship's run must be applied to the first observed altitude as in Ivory's method.

Example.

On March 8th, in lat 54° 20′ N, long 22° 20′ W, the following observations were taken to determine the latitude and longitude by Sumner's method.

Mean Times nearly.	Chron Time.	Obs alts $\underline{\odot}$
8h 27m A.M.	9h 51m 45s	14° 38′ 45″
2 24 P.M.	3 48 45	24 35 37

IE — 1′ 19″, height of the eye 19 feet. Sun's bearing at first observation SE$\frac{3}{4}$E: Ship's run in the interval SbW$\frac{1}{4}$W, 2·5 miles an hour.

On February 20th, at noon, the chronometer was slow on GMT 1m 20s, losing daily 3s.

Note.—(*a*) Find the Greenwich dates nearly, and also those from chronometer.

(*b*) Find the sun's declination and equation of time for each date.

(*c*) Correct the altitudes, and apply the correction for ship's run to the first one.

(*d*) Compute the four longitudes with the two altitudes, latitudes and polar distances.

(*e*) Tabulate them, and obtain latitude and longitude.

(1) Greenwich Date nearly.	Longitude.	(2) Greenwich Date nearly.
d h m s	° ′	d h m s
7 20 27 0	22 20 W	8 2 24 0
+ 1 9 20	4	+ 1 29 20
7 21 36 20	6,0)8,9 20	8 3 53 20
	1 29 20	

(1) Chronometer Greenh Date.	Rate.	(2) Chronometer Greenh Date.
h m s	s	h m s
21 51 45	3 for	3 48 45
Rate + 51	17 days = 51s	Rate + 51
21 52 36		3 49 36
Error + 1 20		Error + 1 20
7 21 53 56		8 3 50 56

Declination for Date (1)	Change.	Equation of Time Date (1)	Change.
° ′ ″	″	m s	s
S 5 1 15·2	—58·40	11 5·68	—·621
—21 19·0	21·9	13·60	21·9
4 39 56·2	52560	10 52·08	5589
90	5840	+ to app T	621
p (1) 94 39 56	11680		1242
	6,0)127,8960		13·5999
	21 19		

Declination for Date (2)	Change.	Equation of Time Date (2)	Change.
S 4° 37′ 51″	−58″·55	10m 50s·57	—·637s
— 3 42	3·8	— 2·42	3·8
4 34 9	46840	10 48·15	5096
90	17565	+ to app T.	1911
p (2) 94 34 9	6,0)222,490		2·4206
	3 42·5		

Correction for Ship's run.

Fig. 26.

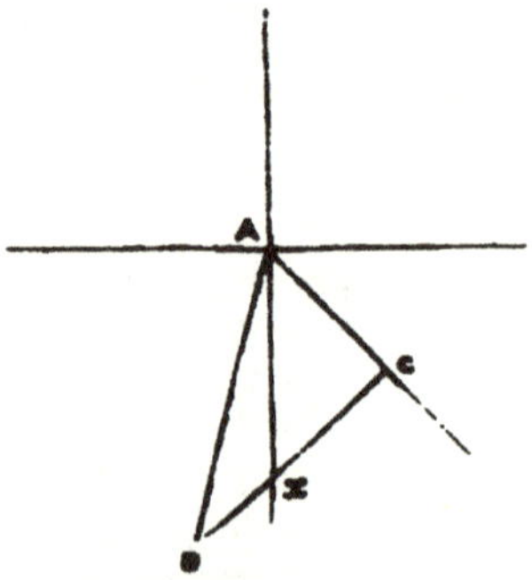

Here $xAC = 4\frac{3}{4}$ pts. being SE $\frac{3}{4}$ E
and $xAB = 1\frac{1}{4}$ " " S b W $\frac{1}{4}$ W
therefore BAC = 6 pts.
and AB = 2·5 × 6 nearly or miles per hour multiplied by number of hours (or interval between chronometer times) = 15 miles.

Log 15′ = 1·164128
Cos 6 pts = 9·582840
Cor 5′·6 = 0·746968

And since the angle BAC is less than 8 points, the correction is additive to the first altitude.

Corrected Altitudes.

First Altitude.

	°	′	″
Obs alt ☉	14	38	45
IE —		1	19
	14	37	26
Dip —		4	17
	14	33	9
Ref —		3	38
	14	29	31
SD +		16	8
	14	45	39
Parallax +			9
	14	45	48
Ship's run corn +		5	36
1st altitude	14	51	24

Second Altitude.

	°	′	″
Obs alt ☉	24	35	37
IE —		1	19
	24	34	18
Dip —		4	17
	24	30	1
Ref —		2	5
	24	27	56
SD +		16	8
	24	44	4
Parallax +			8
2nd altitude	24	44	12

First Observation.

Longitude (a) with lat 54° N.

	° ′ ″	
(1) alt	14 51 24	
(1) lat	54 0 0	sec 10·230781
(1) p	94 39 56	cosec 10·001441
	2)163 31 20	
S	81 45 40	cos 9·156249
a	14 51 24	
(S—a)	66 54 16	sin 9·963718
		2)19·352189
$\frac{1}{2}h$	28 19 0	sin 9·676094
	2	
	56 38 0	
	4	
	60)226 32 0	
	3 46 32	
	24 0 0	
True app T	20 13 28	
Eq T (1)	+ 10 52·1	
True MT	20 24 20·1	
GD (1)	21 53 56·0	
	1 29 35·9	
	60	
	4)89 35 54	
(a)	22° 23′ 58″ W	

Longitude (a') with lat 55° N.

	° ′ ″	
(1) alt	14 51 24	
(2) lat	55 0 0	sec 10·241409
(1) p	94 39 56	cosec 10·001441
	2)164 31 20	
S	82 15 40	cos 9·129215
a	14 51 24	
(S—a)	67 24 16	sin 9·965315
		2)19·337380
$\frac{1}{2}h$	27 47 47	9·668690
	2	
	55 35 34	
	4	
	6,0)22,2 22 16	
	3 42 22·3	
	24 0 0	
True app T	20 17 37·7	
Eq T (1)	+ 10 52·1	
True MT	20 28 29·8	
GD (1)	21 53 56·0	
	1 25 26·2	
	60	
	4)85 26 12	
(a')	21° 21′ 33″ W	

Second Observation.

Longitude (*b*) with latitude 54° N.

(2) alt 24° 44′ 12″
(1) lat 54 0 0 sec 10·230781
(2) *p* 94 34 9 cosec 10·001382
2)173 18 21
S 86 39 10 cos 8·766316
a 24 44 12
(S—*a*) 61 54 58 sin 9·945596
2)18·944075
½*h* 17° 14′ 53″ sin 9·472037
2
34 29 46
4
6,0)13,7 59 4
True app T 2 17 59·1
Eq T (2) + 10 48·1
True MT 2 28 47·2
GD (2) 3 50 56·0
1 22 8·8
60
4)82 8 48
(*b*) 20°32′12″ W

Longitude (*b′*) with latitude 55° N.

(2) alt 24° 44′ 12″
(2) lat 55 0 0 sec 10·241409
(2) *p* 94 34 9 cosec 10.001382
2)174 18 21
S 87 9 10 cos 8·696122
s 24 44 12
(S—*a*) 62 24 58 sin 9·947597
2)18·886510
½*h* 16° 6′ 39″ sin 9·443255
2
32 13 18
4
6,0)12,8 53 12
True app T 2 8 53·2
Eq T (2) + 10 48·1
True MT 2 19 41·3
GD (2) 3 50 56·0
1 31 14·7
60
4)91 14 42
(*b′*) 22°48′40″ W

The four Longitudes are now Tabulated as follows,—

	With lat 54° N.	With lat 55° N.
1st alt 2nd alt	(*a*) 22° 23′ 58″ W (*b*) 20 32 12 W 1 51 46 60 111 60 A 6706 E B 5227 W 11933 (A + B)	(*a′*) 21° 21′ 33″ W (*b′*) 22 48 40 W 1 27 7 60 87 60 B 5227 W

The differences are called E and W because (*b*) is East of (*a*), and (*b′*) West of (*a′*) respectively. A is always the larger difference.

The differences are added because they are of unlike names.

Call the difference of the assumed latitudes, viz., 1° or 60′ = D : then.

$$\text{Cor for lat} = \frac{A \times D}{A + B} = \frac{7706 \times 60'}{11933}$$

Log 7706 =	3·886829
Log 60 =	1·778151
	5·664980
Log 11933 =	4·111699
Correction 35′·75 =	1·553281

The latitude corrected is always that which stands over A; and the above correction is called N or S, as B is north or south of A

Lat (over A)	54°	0′	0″	N
Correction		35	45	N
True latitude	54	35	45	N

$$\text{Cor for long} = \frac{A \times C}{A + B}$$

Long (a)	22°	23′	58″	W
„ (a')	21	21	33	W
	1	2	25	
	60			
	62			
	60			
	3745″ C			

$$\text{Then Cor for long } \frac{7706'' \times 3745''}{11933}$$

Log 7706 =	3·886829
Log 3745 =	3·573452
	7·460281
Log 11933 =	4·111699
6,0)223,1″ =	3·348582
Correction 37′ 11″	

The longitude corrected is always that which stands over A; and the above correction is called E or W, as B is east or west of A.

Long (over A)	22°	23′	58″	W
Correction		37	11	E
True longitude	21	46	47	W

The following diagram may serve to illustrate the conditions of the preceding problem :—

Fig. 27.

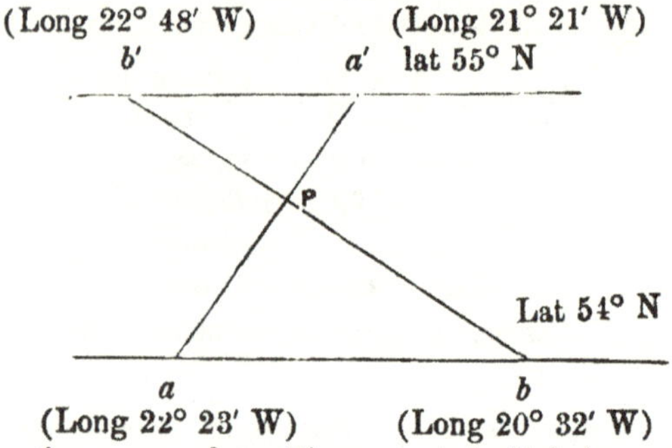

The respective assumed positions *a*, *a′*, *b*, *b′*, being pricked off on a chart, by joining *a* to *a′*, and *b* to *b′* the position P of the ship is obtained.

But the position of the ship may in other cases fall outside the assumed parallels. In the following example, conceive the four longitudes to have been deduced as in the preceding one.

	With lat 40° N	With lat 41° N
	° ′ ″	° ′ ″
1st alt	(*a*) 102 39 30 E	(*a′*) 103 27 30 E
2nd alt	(*b*) 101 40 45 E	(*b′*) 101 40 25 E
	0 58 45 60 58 60 B 3525 W	1 47 5 60 107 60 A 6425 W B 3525 W 2900 (A — B)

$$\text{Cor for lat} = \frac{A \times D}{A - B} = \frac{6425 \times 60}{2900}$$

Log 6425 = 3·807873
Log 60 = 1·778151
5·586024
Log 2900 = 3·462398
6,0)133′ = 2·123626
Correction 2° 13′

	°	′	″	
Lat (over A)	41°	0′	0″	N
Correction	2	13	0	S
True lat	38	47	0	N

$$\begin{array}{lr} (a) & 102^\circ\ 39'\ 30''\ \text{E} \\ (a') & \underline{103\ \ 27\ \ 30\ \ \text{E}} \\ & 0\ \ 48\ \ 0 \\ & \underline{60} \\ & \underline{\underline{2880''}}\ \text{C} \end{array}$$

$$\text{Cor for long} = \frac{\text{A} \times \text{C}}{\text{A} - \text{B}} = \frac{6425'' \times 2880''}{2900}$$

$$\begin{array}{rl} \text{Log } 6425 = & 3{\cdot}807873 \\ \text{Log } 2880 = & \underline{3{\cdot}459392} \\ & 7{\cdot}267265 \\ \text{Log } 2900 = & \underline{3{\cdot}462398} \\ 6{,}0)638{,}0'' = & \underline{\underline{3{\cdot}804867}} \\ \underline{6{,}0)10{,}6'\ 20''} & \\ \text{Correction } \underline{\underline{1^\circ\ 46'\ 20''}} & \end{array}$$

$$\begin{array}{rl} \text{Long (over A)} & 103^\circ\ 27'\ 30''\ \text{E} \\ \text{Correction} & \underline{\ \ \ \ 1\ \ 46\ \ 20\ \ \text{W}} \\ \text{True long} & \underline{\underline{101\ \ 41\ \ 10}}\ \ \text{E} \end{array}$$

The following figure may serve to illustrate the above conditions :—

Fig. 28.

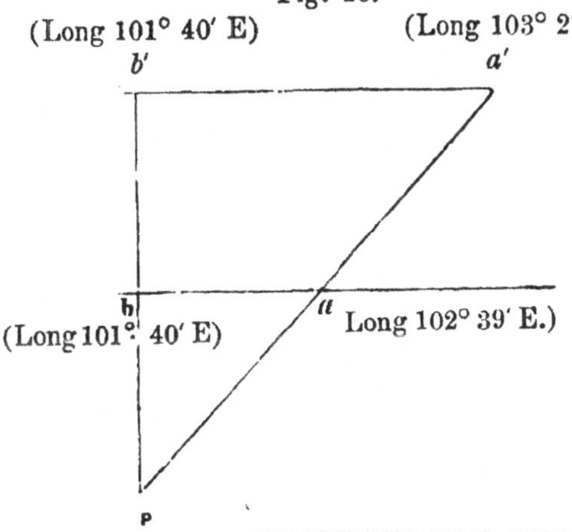

a, *b*, *a′*, *b′*, are the four positions of the ship on the assumed parallels respectively ; then by joining *aa′*, *bb′*, the position P of the ship is found to be below both parallels.

EXAMPLES FOR EXERCISE.

In the following examples the four longitudes are supposed to have been calculated : find the true latitude and longitude.

1.

	with lat 60° N	with lat 61° N
1st alt	(*a*) 49° 30′ 10″ W	(*a′*) 50° 28′ 45″ W
2nd alt	(*b*) 50 27 40 W	(*b′*) 51 12 30 W

Ans. Lat 64° 11′ N ; long 53° 35′ 9″ W.

2.

	with lat 31° S	with lat 31½° S
1st alt	(*a*) 101° 42′ 15″ E	(*a*′) 102° 37′ 10″ E
2nd alt	(*b*) 102 40 15 E	(*b*′) 103 15 30 E

Ans. Lat 32° 28′ 30″ S; long 104° 24′ 12″ E.

LESSER COMPUTATIONS.

(22) To find the Time of a Star's Transit across the Meridian.

Note.—(*a*) From the star's right ascension subtract the sidereal time (increasing the former by 24 hours, if necessary). If this difference be more than 12^h, reckon from the preceding day.

(*b*) To this apply the longitude in time for an approximate Greenwich date.

(*c*) Correct the mean sun's right ascension for this date.

(*d*) From the star's right ascension subtract this corrected mean sun's right ascension, and the result is the time of the star's passage.

Example.

At what time will the star β Leonis pass the meridian of a place in lat 42° 10′ N, long 160° 30′ W, on Jan 10th, 1876, and at what distance N or S of the zenith?

		h	m	s
From Star's R*a*		11	42	44·8
Take Sid time		19	17	27·1
January	9^d	16	25	17·7
Longitude	+	10	42	0·0
GD nearly	10	3	7	17·7

Sidereal Time corrected.

h	m	s
19	17	27·1
		29·6
		1·1
19	17	57·8

		h	m	s
From Star's R*a*		11	42	44·8
Take Mean Sun's R*a*		19	17	57·8
Time of passage	9	16	24	47·0

On January 10^d 4^h 21^m $47^s·0$ A.M.

To find distance N or S of Zenith.

	° ′
Latitude	42 10 N
Declination	15 16 N
Zenith distance	26 54 N

EXAMPLES FOR EXERCISE.

1. At what time will *a* Orionis pass the meridian of lat 15° 12′ N, long 142° 30′ E, on April 9th, 1876, and at what distance N or S of the Zenith?

Ans. Time 4^h 37^m 0^s P.M.
Distance 7° 49′ N

2. At what time will *a* Hydræ pass the meridian of lat 2° 30′ N, long 84° 16′ W, on September 12th, 1876, and at what distance N or S of the Zenith.

Ans. Time 9^h 53^m 36^s A.M.
Distance 10° 37′ S.

(23) Transits of a Number of Bright Stars.

Note.—(*a*) Find the Greenwich Dates for the given times.

(*b*) Find the mean sun's R*a* for each date.

(*c*) Add this to the mean time at place: the result is the R*a* of the meridian, and all stars whose R*a*'s lie between these R*a*'s of the meridian are those that pass.

Example.

What bright Stars will pass the meridian of longitude 72° W, on January 8th, 1876, between the hours of 7 P.M., and 9 P.M.

Greenwich Date (1).	Longitude.	Greenwich Date (2).
January 8^d 7^h 0^m 0^s	72° W	8^d 9^h 0^m 0^s
4 48 0	4	4 48 0
8 11 48 0	6,0)28,8	8 13 48 0
	4^h 48^m	

Mean Sun's R*a* (1).	Mean Sun's R*a* (2).
h m s	h m s
19 9 34·02	19 9 34·02
1 48·42	2 8·13
7·88	7·88
19 11 30·32	19 11 50·03

M.T.P. + Mean Sun's R*a* = R*a* of meridian.

	h m s		h m s
MTP (1)	7 0 0	MTP (2)	9 0 0
Mean Sun's *Ra*	19 11 30	Mean Sun's *Ra*	19 11 50
Ra of mer (1)	2 11 30	*Ra* of mer (2)	4 11 50

Ans. From ξ^2 Ceti to o' Eridani.

EXAMPLE FOR EXERCISE.

1. What bright Stars will pass the meridian of longitude 72° 10′ W, on May 10th, 1876, between the hours of 1 A.M. and 3 A.M.

Ans. From Antares to μ Sagittarii.

2. What bright Stars will pass the meridian of longitude 60° E, on June 20th, 1876, between the hours of 4 P.M. and 6.30 P.M.

Ans. From Regulus to α' Crucis.

(24) To find the Time of Sunrise.

Note.—(*a*) Find the Greenwich Date from a supposed approximate time of sunrise.

(*b*) Correct the Sun's declination for this date.

(*c*) Compute the hour angle at sunset by the formula $\cos h = -\cot p \tan \text{lat}$. Then 12^h — time of sunset = time of rising.

We know that on March 21st, and September 21st, or thereabout, the Sun rises at 6 A.M. and sets at 6 P.M., everywhere. In the Summer half of the year in either hemisphère, he will rise earlier than 6, and in the winter half, later.

Example.

Find the time of Sunrise and Sunset on May 8th, 1876, in latitude 48° 20′ N, longitude 72° 40′ E.

Suppose the Sun to rise at 4 A.M.

Then

	d h m s
Date at place May	7 16 0 0
Longitude	— 4 50 40
G.D. nearly	7 11 9 20

	° ′
Longitude	72 40 E
	4
	6,0)29,0 40
	4 50 40

Declination.		Change.
	° ′ ″	″
N	16 59 25	+ 40·96
	+ 7 35	11· 1
	17 7 0	4096
	90 0 0	4096
p	72 53 0	4096
		6,0)45,4·656
		7·35

Time of Rising and Setting.

$$\text{Cos } h = -\cot p \tan \text{lat.}$$

	Cot p	9·488492
	Tan lat	10·050647
h 69° 45′	cos	9·539139

For properly accounting for the negative sign see investigation of rule: it may be sufficient here to say, that when p is less than 90°, h must be subtracted from 180°.

Then	° 69	′ 45		
	180			
$h =$	110	15		
		4		
	6,0)44,1	0		
Time of Sunset	7	21	0	P.M.
	12	0	0	
Time of Sunrise	4	39	0	A.M.

If it be desired to obtain the time very accurately, find another Greenwich date for which correct the declination, and proceed as before.

EXAMPLES FOR EXERCISE.

1. Find the time of Sunrise and Sunset on July 12th, 1876, in latitude 49° 12′ S, longitude 112° 30′ E.

Ans. Sunrise $7^h\ 51^m\ 28^s$ A.M.
Sunset 4 8 32 P.M.

2. Find the time of Sunrise and Sunset on August 20th, 1876, in latitude 51° 9′ N, longitude 71° 30′ E.

Ans. Sunrise $4^h\ 56^m\ 28^s$ A.M.
Sunset 7 3 32 P.M.

(25) To find the Duration of Twilight.

That parallel of altitude on which the Sun is when twilight commences and ends, is situated 18° below the horizon, and is called the twilight circle. His altitude at that time may be called — 18°.

Example.

It is required to compute the duration of twilight on May 8th, 1876, in latitude 48° 20′ N, longitude 72° 40′ E, (see preceding Example).

	°	′		
a —	18	0		
l	48	20	sec	10·177312
p	72	53	cosec	10·019675
	2)103	13		
S	51	36	cos	9·793195
a —	18	0		
(S−a)	69	36	sin	9·971870
				2)19·962052
	°	′		
$\frac{1}{2}h$	73	11	sin	9·981026
		2		
	146	22		
		4		
	6,0)58,5	28		

Twilight ends	9	45	28
Sun sets at	7	21	0
Duration of twilight	2	24	28

The polar distance at Sunset is taken from the preceding Example.

(26) To find the Time of the Moon's Rising.

If the moon's declination be 0° or nearly 0°, in other words if that body is on the equator, she will rise very nearly 6 hours before coming to the meridian. When her declination is north,—in N latitude her diurnal arc or time above the horizon will be more than 12 hours, and less in S latitude, and *vice versâ*. For a rough computation, we may suppose her to rise 6 hours before coming to the meridian.

Note.—(*a*) Find the Astronomical meridian passage for the day given, from which, subtract 6 hours and apply the longitude.

(*b*) With this approximate Greenwich date correct the moon's declination.

(*c*) Compute the hour angle as in the sun, to which add the meridian passage; the result is the time nearly at which the moon sets. By subtracting the hour angle (called the semi-diurnal arc), from the meridian passage we have the time of rising.

Example.

Find the time of the moon's rising and setting on January 8th, in latitude 45° 20′ N, long 62° 40′ E.

	d	h	m
Mer pass	7	21	7·3
Subtract		6	0·0
Date nearly	7	15	7·3
Longitude —		4	10·7
G. D. nearly	7	10	56·6

	°	′	
Longitude	62	40	E
		4	
	6,0)25,0	40	
	4 10	40	

Declination.	Change.
N 22° 17′ 17″	+ 9″·75
+ 9 16	57
22 26 33	6825
90 0 0	4875
p 67 33 27	6,0)55,5·75
	9′ 16″

Time of Rising.

Cos h = — Cot p tan lat

Cot p	9·615793
Tan lat	10·005053
65° 18′ cos	9·620846
180	
114 42	
4	
6,0)45,8 48	
Semi-diurnal arc	7 38 48

	h	m	s
Semi diurnal arc	7	38	48
Meridian pass	+ 21	7	18
Moon sets January	8	4 46	6
Moon rises January	7	13 28	30

EXAMPLES FOR EXERCISE.

1. When did the Moon rise on July 12th, 1876, in latitude 42° 20′ S, longitude 41° 30′ W.

Ans. Rose at 10^h 23^m 40^s A.M.

2. When did the Moon rise on August 29th, 1876, in latitude 14° 10′ N, longitude 80° E.

Ans. Rose at 2^h 20^m 33^s A.M.

(27) To find the Time of a Star's Rising.

It has been shown that refraction and dip, but especially the former, have a tendency to increase a body's altitude. When a heavenly body is at a low altitude its refraction is greater than when it is nearer the Zenith: the refraction is a maximum (33′), when an object is on the horizon. Hence, a star which is really on the horizon would appear to be 33′ above it. In computing the time of rising of a star, it is necessary to take this into account.

Note.—(a) Add the dip to the refraction, and call this depression (d).

(b) Compute the hour angle with the formula.

$\text{Sin } \frac{1}{2}h = \sqrt{\text{Sin S. cos (S}-d) \text{ sec } l \text{ cosec } p}$, (See Investigation).

(c) To the westerly hour angle add the Star's Ra, and subtract the Sidereal Time, this gives the time of the Star's rising nearly.

(d) Correct the mean Sun's Ra for the Greenwich Date obtained from this, and substitute it for the Sidereal Time.

EXAMPLE.

When did Markab rise on September 20th, 1876, in latitude 51° 20′ N, longitude 2° 30′ W. Height of the eye 24 feet.

```
   Dip for 24 feet =  4′ 49″
               Ref = 33   0
Depression = (d)     37  49
```

Markab's Ra.

22ʰ 58ᵐ 38ˢ·5

Markab's Declination.

```
  N 14° 32′ 40″
    90
p   75  27  20
```

Hour Angle.

$$\text{Sin } \tfrac{1}{2}h = \sqrt{\text{Sin S} . \cos (\text{S}-d) . \sec l \text{ cosec } p}$$

```
            °  ′  ″
(d)         0 37 49
(l)        51 20  0   sec    10·204267
(p)        75 27 20   cosec  10·014146
        2)127 25  9
S          63 42 34   sin     9·952579
(d)         0 37 49
(S−d)      63  4 45   cos     9·655868
                           2)19·826860
            °  ′  ″
½h         55  0 41   sin     9·913430
                  2
          110  1 22
                  4
       6,0)440  5 28
```

```
Easterly Hour Angle          7 20  5·5
                            24  0  0
Westerly Hour Angle         16 39 54·5
Star's Ra                   22 58 38·5
Ra of Meridian              39 38 33·0
Sidereal time on 20th       11 58 52·4
Date nearly September 20ᵈ    3 39 40·6
```

Greenwich Date.

```
               d   h  m  s
Date nearly   20   3 39 41
Longitude        + 10  0
              20   3 49 41
```

Mean Sun's Ra.

```
         h  m  s
        11 58 52·4
For 3ʰ =      29·5
   49ᵐ =       8·0
   41ˢ =        ·1
        11 59 30·0
```

```
                  h  m  s
Ra of Meridian   39 38 33
Mean Sun's Ra    11 59 30
Sep 20ᵈ           3 39  3
```

The Star rose at 3ʰ 39ᵐ P.M. on September 20th, 1876.

EXAMPLES FOR EXERCISE.

1. When did Altair rise in lat 51° 27′ N, long 40° E, on September 10th, 1876 : height of eye 20 feet.

Ans. At 1^h 37^m 54^s P.M.

2. When did *a* Andromeda rise in lat 59° 30′ N, long 76° W, on October 12th, 1876 : height of eye 27 feet.

Ans. At 11^h 53^m 12^s A.M.

(28) To find the Time of High Water.

TABLE.

Time of Moon's Mer Passage.		Moon's Semi-diameter.								
		14′30″	14′45″	15′ 0″	15′15″	15′30″	15′45″	16′ 0″	16′15″	16′30″
h m	h m	h m	h m	h m	h m	h m	h m	h m	h m	h m
0 0	12 0	—0 4	—0 3	—0 2	—0 1	—0 0	+0 1	+0 2	+0 3	+0 5
0 20	12 20	—0 8	—0 8	—0 7	—0 6	—0 5	—0. 4	—0 3	—0 2	—0 1
0 40	12 40	—0 12	—0 12	—0 11	—0 11	—0 10	—0 10	—0 9	—0 9	—0 8
1 0	13 0	—0 17	—0 17	—0 17	—0 16	—0 16	—0 16	—0 15	—0 15	—0 15
1 20	13 20	—0 22	—0 22	—0 22	—0 22	—0 22	—0 22	—0 22	—0 22	—0 22
1 40	13 40	—0 27	—0 27	—0 27	—0 28	—0 28	—0 28	—0 29	—0 29	—0 29
2 0	14 0	—0 31	—0 31	—0 32	—0 32	—0 33	—0 33	—0 34	—0 35	—0 36
2 20	14 20	—0 36	—0 36	—0 37	—0 38	—0 39	—0 40	—0 41	—0 42	—0 43
2 40	14 40	—0 40	—0 41	—0 42	—0 43	—0 44	—0 45	—0 46	—0 47	—0 49
3 0	15 0	—0 44	—0 45	—0 47	—0 48	—0 49	—0 51	—0 52	—0 54	—0 56
3 20	15 20	—0 48	—0 49	—0 51	—0 52	—0 53	—0 55	—0 57	—0 59	—1 1
3 40	15 40	—0 51	—0 53	—0 55	—0 57	—0 58	—1 0	—1 2	—1 4	—1 7
4 0	16 0	—0 55	—0 57	—0 59	—1 0	—1 2	—1 5	—1 7	—1 10	—1 12
4 20	16 20	—0 57	—0 59	—1 0	—1 2	—1 4	—1 6	—1 9	—1 12	—1 15
4 40	16 40	—0 59	—1 1	—1 3	—1 5	—1 7	—1 9	—1 12	—1 15	—1 18
5 0	17 0	—1 0	—1 2	—1 4	—1 6	—1 8	—1 10	—1 13	—1 16	—1 19
5 20	17 20	—1 0	—1 2	—1 4	—1 6	—1 8	—1 11	—1 14	—1 17	—1 20
5 40	17 40	—0 58	—1 0	—1 2	—1 4	—1 6	—1 8	—1 11	—1 13	—1 16
6 0	18 0	—0 55	—0 56	—0 58	—1 0	—1 2	—1 4	—1 6	—1 9	—1 12
6 20	18 20	—0 49	—0 50	—0 51	—0 52	—0 54	—0 56	—0 58	—1 0	—1 3
6 40	18 40	—0 43	—0 44	—0 45	—0 46	—0 47	—0 48	—0 49	—0 51	—0 53
6 50	18 50	—0 37	—0 37	—0 38	—0 39	—0 40	—0 41	—0 42	—0 43	—0 45
7 0	19 0	—0 32	—0 32	—0 33	—0 33	—0 34	—0 34	—0 35	—0 36	—0 37
7 10	19 10	—0 27	—0 27	—0 27	—0 27	—0 28	—0 28	—0 28	—0 29	—0 29
7 20	19 20	—0 22	—0 22	—0 22	—0 22	—0 22	—0 22	—0 22	—0 22	—0 22
7 30	19 30	—0 16	—0 16	—0 16	—0 15	—0 15	—0 15	—0 14	—0 14	—0 14
7 40	19 40	—0 11	—0 11	—0 10	—0 10	—0 9	—0 9	—0 8	—0 7	—0 6
7 50	19 50	—0 6	—0 6	—0 5	—0 4	—0 3	—0 2	—0 1	—0 0	+0 2
8 0	20 0	—0 1	—0 0	+0 1	+0 2	+0 3	+0 4	+0 6	+0 7	+0 9
8 20	20 20	+0 5	+0 6	+0 7	+0 9	+0 11	+0 13	+0 15	+0 17	+0 19
8 40	20 40	+0 11	+0 12	+0 14	+0 16	+0 18	+0 20	+0 22	+0 25	+0 28
9 0	21 0	+0 14	+0 16	+0 18	+0 20	+0 22	+0 24	+0 26	+0 29	+0 32
9 20	21 20	+0 16	+0 18	+0 20	+0 22	+0 24	+0 27	+0 33	+0 33	+0 36
10 0	22 0	+0 15	+0 17	+0 19	+0 21	+0 23	+0 25	+0 27	+0 30	+0 34
10 20	22 20	+0 13	+0 15	+0 17	+0 19	+0 21	+0 23	+0 25	+0 28	+0 31
10 40	22 40	+0 11	+0 13	+0 14	+0 16	+0 18	+0 20	+0 22	+0 25	+0 28
11 0	23 0	+0 7	+0 8	+0 10	+0 12	+0 14	+0 16	+0 18	+0 20	+0 23
11 20	23 20	+0 4	+0 5	+0 6	+0 7	+0 9	+0 11	+0 13	+0 15	+0 17
11 40	23 40	+0 0	+0 1	+0 2	+0 3	+0 5	+0 7	+0 8	+0 10	+0 12
12 0	24 0	—0 5	—0 3	—0 2	—0 1	—0 0	+0 1	+0 2	+0 3	+0 5

Note.—Take out moon's meridian passage for the given day: add to this the "Establishment of the Port," *i.e.*, the time of high water at full and change of the moon: then apply correction from the preceding table.

EXAMPLE.

1. Find the time of high water at London Bridge on January 17th, 1876.

		h m
Moon's meridian passage		5 24·5
Establishment of Port Page 488 Nautical Almanac 1876	+	1 58·0
		7 22.5
Correction (see table)	—	1 6
		6 16·5

Therefore the time of high water is 6^h 16^m P.M.
And the A.M. tide is 12^h 24^m earlier nearly.
i.e. 5^h 52^m A.M.

EXAMPLES FOR EXERCISE.

If after applying the Establishment and Correction, the sum exceeds 12^h 24^m or 24^h 48^m, subtract those quantities, and the result is the P.M. tide for the day.

1. Find the time of high water at Hull on May 10th, 1876.
Ans. 6^h 58^m P.M.

2. Find the time of high water at Liverpool on July 12th, 1876.
Ans. 2^h 24^m P.M.

3. Find the time of high water at Newcastle on December 29th, 1876.
Ans. 2^h 34^m P.M.

VARIATION OF THE COMPASS.

There are two Astronomical methods of determining this, first by observed Amplitudes, and secondly, by Azimuths of the sun or some other heavenly body. The sum or difference between any observed Azimuth or Amplitude, and that deduced by computation is the error of the Compass, including both Variation and Deviation. If the variation alone be required the deviation must be applied to the compass error, in order to determine it.

(29) VARIATION BY AMPLITUDE.

Amplitudes are measured from the east or west, according to the position of a body on the horizon,—whether rising or setting; and as the body must travel to the north of the equator if it have north declination, and to the south if south, it follows that it will rise on that side of the east point or set on that of the west which is of the same name as the declination.

EXAMPLE.

On January 12th, 1876, at about 4^h 30^m A.M., mean time nearly, in lat 39° 20′ S, long 36° 45′ W, the sun rose by compass E 19° 33′ S : deviation 2° 30′ E, find the variation of the compass.

Note.—(*a*) Find the Greenwich date.

(*b*) Correct the declination.

(*c*) Compute the true amplitude by the formula, Sin amp = sin dec . sec lat.

(*d*) This is named as explained above: then by applying the observed amplitude, the sum is the variation if one be N and the other S; the difference if alike. The variation is east or west according as the true bearing is right or left of the observed.

Greenwich Date.	Longitude.	Declination.	Change.
d h m s	° ′	° ′ ″	″
11 16 30 0	36 45 W	S 21 51 58	— 23·14
+ 2 27 0	4	— 7 20	19
11 18 57 0	6,0)147 0	21 44 38	20826
	+ 2 27·0		2314
			6,0)43,9·66
			7·20

True Amplitude.

Sin amp = Sin dec Sec lat.
Sin dec = 9·568539
Sec lat = 10·111556
True amp 28° 36′ sin 9·680095

Obs amp E 19° 33′ S *r**		True amp E 28° 36′ S	
Dev 2 30 E *r*		Comp error E 22 3 S	
Comp error E 22 3 S *r*		Variation 6 33 E	

Fig. 29.

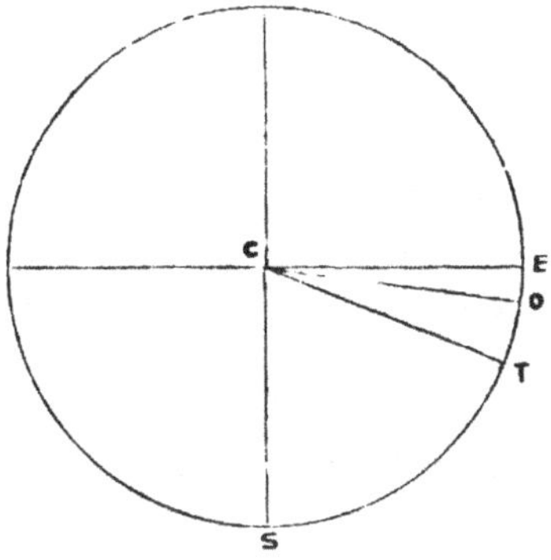

Here measuring from E to S, the true amp viz ET = 28°, while the observed viz EO = 22°. Hence as T is on the right of O as seen from C, the variation is East. The observed amplitude is cleared of deviation as a compass Course to find the True Course.

* Called *r* because South is right of East.

EXAMPLES FOR EXERCISE.

1. February 20th, 1876, at about 7^h 30^m A.M., mean time in latitude 40° 26′ N, longitude 80° 45′ E, the Sun rose by compass E 15° 40′ N: deviation 2° 10′ W: find the variation.

Ans. 32° 36′ E.

2. March 20th, 1876, at about 6^h A.M., mean time in latitude 54° 20′ N, longitude 30° 44′ E, the Sun rose by compass E 5° S: deviation 2° 15′ W: find the variation.

Ans. 2° 41′ W.

ELEMENTS FOR PRECEDING EXAMPLES FROM NAUTICAL ALMANAC, 1876.

1. Sun's declination S 11° 23′ 53″ : change — 53″·29.
2. Sun's declination S 0° 17′ 57″ : change — 59″·27.

(30) VARIATION WITH THE ALTITUDE OR AZIMUTH ALTITUDE.

The Azimuth of a body is its bearing when in altitude, and for conventionality, is reckoned **contrary to the latitude**, (see this explained in Investigation of method). It is also reckoned East or West as the body is **East or West of the meridian of the place of observation.**

EXAMPLE.

On January 12th, 1876, at about 9^h 30^m A.M., mean time, in latitude 39° 50′ S, longitude 36° 45′ W, the obs alt ☉ was 51° 50′ 10″, IE + 1′ 15″: Height of the eye 23 feet: Sun's bearing at same time N 50° 16′ E, deviation 2° E: find the variation of the compass.

Note.—(*a*) Find the Greenwich Date.

(*b*) Correct the declination and find the polar distance (*p*).

(*c*) Correct the altitude (*a*).

(*d*) Compute the true Azimuth from the formula

$$\text{Sin } \tfrac{1}{2} \text{ Az} = \sqrt{\text{Sec } a \text{ sec } l \cos \text{S} . \cos (\text{S}-p)}$$

(*e*) Apply observed bearing as above.

Greenwich Date.	Longitude.	Declination.	Change.
d h m s	° ′	° ′ ″	″
11 21 30 0	36 45 W	S 21 51 58	— 23·14
2 27 0	4	— 9 13	23· 9
11 23 57 0	6,0)14,7 0	21 42 45	20826
	2 27 0	90 0 0	6942
		p 68 17 15	4628
			6,0)553·046
			9·13

True Altitude.

	° ′ ″
Obs alt ☉	51 50 10
IE +	1 15
	51 51 25
Dep —	4 43
	51 46 42
Ref —	45
	51 45 57
SD +	16 18
	52 2 15
Parallax +	5
(a)	52 2 20

True Azimuth.

$$\text{Sin}\ \tfrac{1}{2}\text{Az} = \sqrt{\text{Sec}\ a\ \text{sec}\ l\ \cos S\ \cos(S-p)}$$

	° ′		
(a)	52 2	sec	10·210982
(l)	39 50	sec	10·114689
(p)	68 17		
	2)160 9		
S	80 4	cos	9·236795
p	68 17		
(S—p)	11 47	cos	9·990750
			2)19·553216
	36 43	sin	9·776608
	2		
True Az N	73 26 E		

	° ′	
Obs Azimuth N	50 16	E(r)
Deviation	2 0	E(r)
Comp Error N	52 16	E

	° ′	
True Azimuth N	73 26	Er
Comp Error N	52 16	E
Variation	21 10	E

Fig. 30.

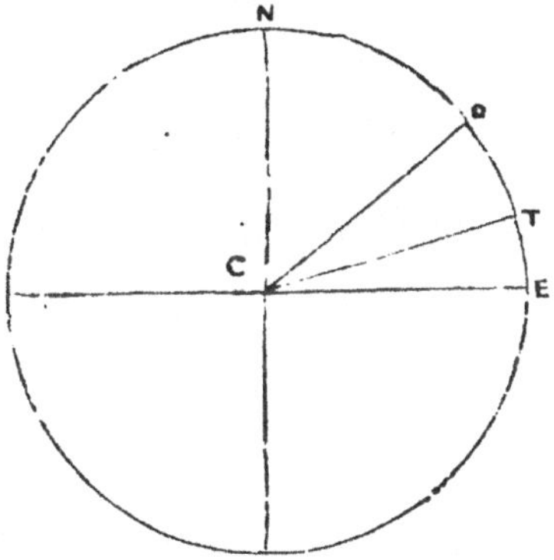

Here, as before, by measuring from the North, T is on the right of O : hence the Variation is East.

Here it is necessary to notice, in order to guard against errors in future, that if the observed bearing and true bearing be of contrary names, one of them, (generally the observed), must be subtracted from 180° and the name reversed, thus :—

Given the true bearing N 110° 30′E, and the observed S 62° 30′ E : then,

	° ′	
S	62 30	E
	180	
N	117 30	E

	° ′	
True Azimuth N	110 10	E
Obs — N	117 30	E
Comp Error	7 20	W

EXAMPLES FOR EXERCISE.

1. January 27th, 1876, at $3^h\ 45^m$ P.M., mean time in latitude 24° 18′ N, longitude 106° 30′ W, the obs alt ☉ was 22° 20′ 10″, IE + 1′ 12″ : height of the eye 17 feet : Sun's bearing S 50° 45′ W, deviation 1° 30′ E, find the variation of the compass.

Ans. 3° 33′ E.

2. February 12th, 1876, at 7^h 45^m A.M., mean time in latitude 10° 42′ S, longitude 18° 40 W, the obs alt ☉ was 24° 42′ 0′, IE + 1′ 15′: height of the eye 12 feet: Sun's bearing N 84° 20′ E, deviation 1° 10′ W, find the variation of the compass.

Ans. 17° 16′ E.

ELEMENTS FOR PRECEDING EXAMPLES FROM NAUTICAL ALMANAC, 1876.

1. Sun's declination S 18° 32′ 53″ : Change — 38″·50 : Semi diam 16′ 17″.
2. Sun's declination S 14° 8′ 4″ : Change — 49″·14 : Semi diam 16′ 14″.

(31) VARIATION WITHOUT THE ALTITUDE OR AZIMUTH HOUR ANGLE.

This is a method seldom, if ever, used at Sea, inasmuch as the exact time must be known, and the calculation is longer than those of the other methods. It may be useful, however, for verifying other results. The time Azimuth is obtained by the direct rules of Spherical Trigonometry, (see Investigation of Method), and the observed is applied as before.

EXAMPLE 1.

February 19th, 1876, at 10^h 30^m 30^s A.M., true mean time at place, latitude 51° 12′ N, longitude 31° 42′ W, the Sun bore by compass S 13° 49′ E, deviation 1° 10′ W : find the variation of the compass.

Note.—(*a*) Find the Greenwich Date.

(*b*) Take out the declination, and, (if the time in question be mean time), the equation of time also.

(*c*) Find the Hour Angle : (see subordinate computation H).

(*d*) Find the Colatitude.

(*e*) Then (see figure) $\text{Tan}\ \frac{1}{2}(Z+S) = \text{Cos}\ \frac{p-l'}{2}\ \text{sec}\ \frac{p+l'}{2}\ \text{cot}\ \frac{h}{2}$

$$\text{And Tan}\ \frac{1}{2}(Z-S) = \text{Sin}\ \frac{p-l'}{2}\ \text{cosec}\ \frac{p+l'}{2}\ \text{cot}\ \frac{h}{2}$$

$\frac{1}{2}(Z+S)+\frac{1}{2}(Z-S)$ is the Azimuth when p is greater than l': $\frac{1}{2}(Z+S)$ — $\frac{1}{2}(Z-S)$, when it is less : (for explanation see investigation).

Greenwich Date.	Longitude.	Declination.	Change.
d h m s	° ′	° ′ ″	″
18 22 30 30	31 42 W	S 11 23 53	— 53·29
+ 2 6 48	4	— 32	·6
19 0 37 18	6,0)12,6 48	11 23 21	31·974
	2 6 48	90	
		(*p*) 101 23 21	

Equation of Time.	Change.	Colatitude.
m s	″	° ′
14 6·70	—·235	Lat 51 12 N
— ·14	·6	90 0
14 6·56	·1410	*l′* 38 48
— To Mean Time		

Hour Angle.

	h	m	s
Mean Time	22	30	30
Eq of Time	—	14	6
	22	16	24
	24	0	0
2)	1	43	36
	0	51	48
		60	
4)		51	48
$\frac{1}{2}h$		12°	57′

True Azimuth.

$$\text{Tan}\ \frac{Z+S}{2} = \cos\frac{p-l'}{2}\ \sec\frac{p+l'}{2}\ \cot\frac{h}{2}$$

$$\text{Tan}\ \frac{Z-S}{2} = \sin\frac{p-l'}{2}\ \text{cosec}\frac{p+l'}{2}\ \cot\frac{h}{2}$$

	° ′				
$\frac{h}{2}$	12 57	cot	10·638368	cot	10·638368
p	101 23				
l'	38 48				
2)	62 35				
$\frac{1}{2}(p-l')$	31 17	cos	9·931768	sin	9·715394
2)	140 11				
$\frac{1}{2}(p+l')$	70 5	sec	10·467688	cosec	10·026785
$\frac{1}{2}$(Z + S)	84 46	tan	11·037824	tan	10·380547
$\frac{1}{2}$(Z − S)	67 24				
	152 10				
	180 0				
True Az:	S 27 50 E				

Obs Az:	S 13° 49′ E (l)		True Az:	S 27° 50′ E
Dev:	1 10 W (l)		Comp Error	S 14 59 E
Comp Error	S 14 59 E		Variation	12 51 W

The figure is laid off as before.

Example 2.

February 12th, 1876, at 8h 40m 40s P.M., mean time, in lat 65° 50′ N, long 82° 30′ E, the star Capella bore S 42° 10′ W: deviation 2° 30′ W: find the variation of the compass.

Note.—(*a*) Find the Greenwich date.

(*b*) Take out the star's R*a* and declination, and the mean sun's R*a*.

(*c*) Hour angle = MTP + mean sun's R*a* — Stars R*a* (see subordinate computation H).

(*d*) Computation as above.

Greenwich date.	Longitude.	Star's R*a* and dec.	Mean Sun's R*a*.
d h m s	° ′	h m s	h m s
12 8 40 40	82 30 E	5 7 33·2	21 27 33·5
— 5 30 0	4	N 45 52 27	29·6
12 3 10 40	6,0)33,0 0	p 44 7 33	1·6
	5 30 0		·1
			21 28 4·8

Hour Angle.			Computation.				
	h m s			° ′			
MTP	8 40 40		½ h	7 39	cot 10·871870	cot	10·871870
Mean Sun's Ra	21 28 5		p	44 7			
	30 8 45		l′	24 10			
Star's Ra	5 7 33			2)19 57			
	2) 1 1 12		$\frac{p-l'}{2}$	9 58	cos 9·993396	sin	9·238235
	0 30 36						
	60			2)68 17			
	4)30 36 0		$\frac{p+l'}{2}$	34 8	sec 10·082109	cosec	10·250944
½ h	7 39 0						
			$\frac{Z+S}{2}$	83 33	tan 10·947375	tan	10·361049
			$\frac{Z-S}{2}$	66 28			
				150 1			
				180 0			
			True Az S	29 59	W (Star West of Meridian)		

	° ′			° ′	
Obs Az S	42 10	W (r)	True Az S	29 59	W
Deviation	2 30	W (l)	Comp error S	39 40	W
Comp error S	39 40	W	Variation	9 41	W

EXAMPLES FOR EXERCISE.

1. January 12th, 1876, at 9h 34m 40s A.M., apparent time, in latitude 42° 40′ N, longitude 37° 16′ W, the Sun bore by compass S 20° E: deviation 1° 50′ E: find the variation of the Compass.

Ans. 17° 5′ W.

2. April 20th, 1876, at 4h 20m 30s P.M., mean time, in latitude 49° 30′ N, longitude 76° 40′ E, the Sun bore by compass S 72° 10′ W: deviation 4° 30′ E: find the variation of the Compass.

Ans. 2° 3′ E.

ELEMENTS FOR PRECEDING EXAMPLES FROM NAUTICAL ALMANAC, 1876.

1. Sun's declination S 21° 42′ 30″: Change − 24″·19.
2. Sun's declination N 11° 23′ 35″: Change + 51″·59.

Equation of Time 1m 1s·87: Change + ·543, + to Mean Time.

THE CHRONOMETER.

A chronometer is a watch of superior construction, furnished with an expansion balance composed of metals of different expansive qualities, as brass and steel, by which arrangement compensation is made for various changes of temperature, which are found seriously to interfere with the rates of common watches.

This instrument is suspended by gimbals in a box, so as to be unaffected by any motion of the ship, and in vessels where great attention is paid to them they are generally kept in a small apartment called the chronometer room, the temperature of which may be regulated by lamps, and they are further placed in cases lined with soft cushions, so as to be defended from the vibrations of the ship. Several important facts connected with the management of chronometers, may be noted,—the principal of which are as follows:—

When a chronometer is placed in position it should not be moved : it has been found that by reversing the direction of the figures on the face, its rate has been altered. But when necessary to move it, care should be taken to compare it with another standard, both before moving it and when it is again placed in position, in order to ascertain whether its rate has been disturbed.

In winding up a chronometer which has run down, care should be taken to give it a few half turns in a circular direction when it will resume its work : the rate will generally be found to have altered from the former one : and it is necessary to wind up the chronometer at regular intervals, in order to preserve the rate.

The time at ship is kept by a separate instrument termed the **hack watch**, by the aid of which, observations of the heavenly bodies are taken. The chronometer shows **Greenwich Mean Time** alone, and with this all other time is compared. Before leaving home the rate of the chronometer and its error on Greenwich Mean Time are ascertained, from which the time at Greenwich may subsequently be known ; and if there is reason to suspect that the chronometer has changed its rate, the new rate may be found by lunar observations, as previously explained.

It is usual for careful navigators to compare their chronometers when going down the Channel by good observations taken when passing, or in sight of some well known headland or position. It will then be seen whether the error allowed be correct or not.

THE SEXTANT.

The Sextant is the instrument of cœlo-navigation or Nautical Astronomy and is used for measuring angles in any plane. At sea, its use is confined chiefly to the measurements of altitudes and lunar distances. The arc contains the sixth part of a circle, hence the name of the instrument. A similar but rougher appliance whose limb is one-eighth of a circle or **octant**, is familiarly known as a quadrant.

Attached to the framework are two mirrors named the **Index** and **Horizon glasses**, the former being fixed on a moveable radius, so that its inclination to the other may be altered, and the reflected image seen in it, may be again reflected to the horizon-glass : thus, two images of any object may be seen in the latter, one by direct vision, the other by reflection. The extremity of the moveable radius traverses the arc which is called the **limb**, and on which are cut on inlaid plates of platinum divisions consisting of degrees, sub-divided into others of 10′ each. The end of the radius is furnished with a smaller scale called the **Vernier**, so named from its inventor ; this is divided into minutes and seconds by which the overplus divisions not given on the limb may be read off. The vernier is furnished with two screws, one called a **clamp-screw**, at its back, which, by means of a spring, secures it to the limb when an angle is to be taken : the other, a delicate contrivance, named the **tangent-screw**, by which the vernier may be moved backwards or forwards, very **slowly**, while that part of the sextant is prevented from free motion by the clamp-screw. The design of this is, that a contact between any two objects may be slowly and deliberately secured. The vernier is also furnished with a microscope for the purpose of reading its divisions more readily.

The Sextant is furnished with **shades**, which can be made to cover either of the glasses so as to moderate the brightness of any image, and also telescopes, which are made to fit through a collar and point towards the horizon-glass. One of these, the longest, is named the **Inverting Telescope**, more powerful than the others, and containing in its focus two sets of parallel wires forming a centre square, into which the images observed are to be brought. There is also a large screw at the back of the instrument, by which the telescope collar may be raised or lowered, to make the **line of collimation** pass through the centre of the horizon-glass. This collar consists of a double ring the parts of which are joined by screws for the purposes of collimation adjustment. The index and horizon glasses are also furnished with screws, by which they may be adjusted

Fig. 31.

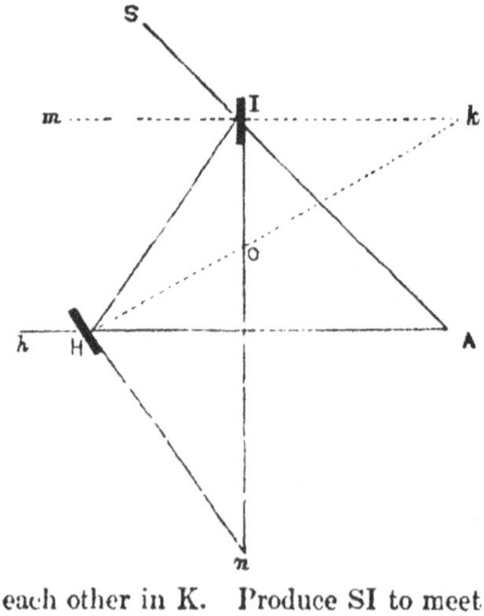

Let I and H represent, respectively, the index and horizon glasses, and S the place of a heavenly body. Then a ray of light SI, from S, will strike the surface of the glass I and be reflected off again at the same angle at which it struck the glass, or by the well known law of optics, the angle of **incidence** is equal to the angle of **reflection**. It follows, therefore, that $SIm = mIH$, and the image of S will be seen at H. Let A H*h* be a horizontal line drawn from the observer's eye at A. Let *m*K and HK be perpendiculars to the surfaces of the glasses, intersecting each other in K. Produce SI to meet AH in A ; and the directions of the glasses I and H to meet in N. Then SAH is the altitude of any heavenly body at S, and N measures the inclination of the glasses.

Then in the triangles IOK, NOH

The angle IOK = HON
And OIK = OHN
Therefore K = N.

Again in the triangle IHA with its side produced to S

SIH = IHA + A
Therefore $\frac{1}{2}$SIH = $\frac{1}{2}$IHA + $\frac{1}{2}$A
But $\frac{1}{2}$SIH = mIH
Therefore mIH = $\frac{1}{2}$IHA + $\frac{1}{2}$A.

But, in the triangle IHK with its side produced to m

The angle mIH = IHK + K
But mIH = $\frac{1}{2}$IHA + $\frac{1}{2}$A
Hence $\frac{1}{2}$IHA + $\frac{1}{2}$A = IHK + K
But $\frac{1}{2}$IHA = IHK
Therefore $\frac{1}{2}$A = K
But K = N
Therefore N = $\frac{1}{2}$A.

Hence, the inclination of the glasses is the measure of half the altitude of a body.

But as it is not the inclination of the glasses, but the altitude that is required, the half degrees are marked as whole degrees on the limb of the instrument, so that the arc read off may be the measure of the angle observed.

The vernier is constructed thus. If 120 divisions of the limb were taken as the vernier scale, then whenever any one division coincided with a division on the limb the others would do the same. But by taking 119 divisions of the limb and sub-dividing into 60 equal parts, only one line of the vernier scale can coincide at a time with any one division of the limb. Thus, the extremities of the vernier marked 0 and 10 will coincide together, but if these be thrown out of coincidence, some other line between zero and the tenth minute must coincide. By this arrangement the odd minutes and seconds from the limb are read off, and each division on the vernier scale is 10″ shorter than any one division of the limb.

Let n represent a division on the limb ; m on the vernier.
Then mn = value of these divisions.

Divide this into $(m + 1)$ equal parts, then, $\frac{mn}{m + 1}$ = a division on the Vernier.

Therefore $n - \frac{mn}{m + 1} = \frac{n}{m + 1}$ the difference between a division of the limb and one of the vernier.

If $n = 10'$, and we wish to divide into $10''$,

$$\frac{n}{m+1} = 10'', \text{ or } \frac{10'}{m+1} = 10'', \text{ or } \frac{600''}{m+1} = 10''$$

Whence $m = 59$.

Thus 119 divisions of the limb are taken for the vernier scale.

The Sextant is liable to certain imperfections and errors which must be duly adjusted. The absence of these is tested thus—

Imperfections.—(*a*) The joints of the instrument must be close and tight, and the screws must work well. (*b*) The inlaid plates containing the divisions of the limb must be on a level with the plane of the sextant. (*c*) The spaces containing the degrees and minutes and seconds must be equal, each to each. (*d*) The glasses of the reflectors should each have their two faces ground and polished parallel to one another. (*e*) The opposite faces of any shade should be parallel.

Methods of detection.—(*b*) (*c*) Bring the index on the vernier into successive coincidence with each division of the limb till the last on the vernier reaches the last on the limb : if in every case, they do not coincide, the instrument is badly graduated.

(*d*) Look with a small telescope into each reflector, separately, in a very oblique direction, and observe the image of a distant object; the image should be **clear** and **distinct** in every part of the glass, and **well defined** about the edges.

(*e*) Fit on a dark glass to the end of the telescope and take an altitude without the shades. Then removing the dark glass, observe, whether, with each shade separately up, the images remain in contact. By this means, **shade error** may be determined, and also noted.

ADJUSTMENTS.

When a sextant is in adjustment and perfect order, both the index and horizon glasses should be **perpendicular to the plane of the instrument**: the parts of the telescope collar should be parallel to one another.

1. To make the index glass perpendicular to the plane of the instrument. Move the index on the vernier to about 60° on the limb, then look obliquely into the glass. The reflected image of the limb as seen in it should be in an exact line with the limb itself. Should one be above the other this glass is not perpendicular to its plane.

Means of Adjustment.—Screws at the back of the glass, by which it is moved forwards or backwards.

2. To make the horizon glass perpendicular to the plane of the instrument. Set the index on the vernier at zero, and screw in the telescope. The reflected and direct images of the sun or other body ought then to coincide. Should one be **to the side of the other**, this adjustment is not in order.

Means of Adjustment.—A capstan screw at the head of the horizon glass moved by a small pin to be found in the box.

3. To make the index and horizon glasses parallel, the index on the Vernier being at zero, the reflected and direct images of the sun, or other body, as seen through the telescope, should coincide with one another. If one be above the other, this adjustment is out of order.

Means of Adjustment.—Capstan screw at the bottom of the horizon glass.

The latter adjustment is not of so much importance as the other two, inasmuch as the index error being determined, it matters not whether the glasses are parallel when the index is at 0. When all these adjustments are made, the reflected image of an object as seen in the horizon glass, should coincide with the direct image.

Index Error.

If the reflected and the direct images do not coincide when the index is at zero, there is an index error.

To determine it.

1. The images may be made to coincide by means of the tangent screw. The arc read off from 0 is the index error, + if the index be off the arc, and — if on. Or,
2. A contact is made with the upper and lower limbs of the sun successively, *i.e.*, the reflected image is brought into contact with the other on opposite sides of it. A quarter of the sum of these readings is the semi-diameter for the day as contained in the almanac, and half their difference is the index error, + if the off reading be greater than the on, — if the reverse. This method is the one generally adopted. A series of observations on each limb is preferable to one only.

Care in the use of a Sextant.

1. The instrument after being used, should never be left lying in the sun, as the glasses will thereby be injured.
2. The various powers of the telescope should be practised with, as a good observation often depends upon using a suitable power.
3. In taking an altitude of the sun in the morning, the best contact is obtained by setting the Vernier in advance, and awaiting the observation: the reverse in the afternoon.
4. The same rule applies to lunar distances.
5. The screws and tubes, &c., should be used with care: the instrument should always be held by the handle: any rough usage soon tells upon the value of the Sextant.

ERROR OF COLLIMATION.

The telescope collar consists of two parts joined by two screws. If one of these screws be fastened more tightly than the other, the parts of the collar will not be parallel, and the axis of the telescope itself, when fixed, will not be parallel to the plane of the instrument. In order to test this, the sun and moon, as suitable objects, are brought into contact on one of the wires in the focus of the telescope; then, if, after shifting the images over the opposite and parallel wire, they are not in contact, either of the screws in the telescope collar should be tightened or loosened, until the adjustments are completed. When once made, it is seldom liable to change.

PART III.

CONTAINING

INVESTIGATIONS OF RULES.

INVESTIGATIONS

OF THE

RULES IN NAVIGATION & NAUTICAL ASTRONOMY.

Parallel Sailing.

Fig. 32.

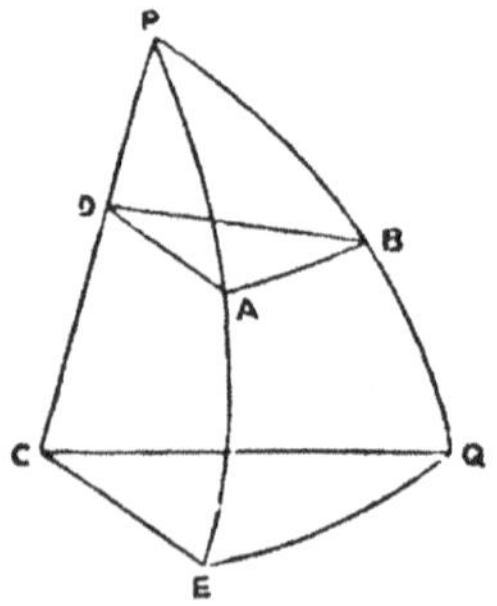

Let A and B be two places upon the same parallel of latitude AB; PE, PQ the meridians passing over those places; EQ the corresponding arc of the Equator. Let PC represent the polar axis of the earth; C the centre of the earth; D the centre of the circle of which AB is an arc. Join DA, DB; CE, CQ.

Then AB is the meridian distance between A and B: EQ is the difference of longitude, and AE or BQ measures the latitude of the parallel AB.

Now, since similar arcs of circles are to each other as their radii, therefore,

$$AB : EQ :: DA : CE$$

$$\text{or mer dist} : \text{D long} :: DA : CE$$

But DA is the sine of the arc AP

Similarly, CE „ „ „ EP

Therefore,

$$\text{Mer dist} : \text{D long} :: \sin AP : \sin EP$$

$$\text{but } \sin AP = \cos AE = \cos \text{lat}$$

$$\text{and } \sin EP = \sin 90° = 1.$$

Therefore,

$$\text{Mer dist} : \text{D long} :: \cos \text{lat} : 1.$$

Hence,

$$\text{Mer dist} = \text{D long} \cos \text{lat} \quad \text{(I)}.$$

$$\text{or } \frac{\text{mer dist}}{\text{D long}} = \cos \text{lat} \quad . \quad . \quad \text{(II)}.$$

$$\text{and D long} = \text{mer dist} \sec \text{lat} \quad \text{(III)}.$$

N

Middle Latitude Sailing.

Fig. 33.

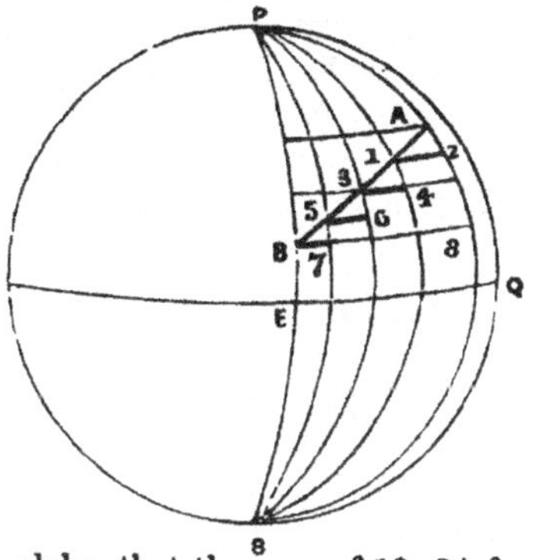

This Sailing is founded upon the supposition, that. the arc of the parallel of middle latitude between two places is nearly equal to their departure. It is, in fact, a special case of parallel sailing. If the ship be supposed to sail from A to B, then AB is the rhumb line, and the arcs 12, 34 &c., are the meridian distances between A and B. It may be shown, experimentally, upon a globe, that the sum of 12, 34 &c., is not equal in length to B8 as might be supposed from the figure, or as would actually be the case upon a plane chart, but is really equal to the arc of that parallel which lies midway between A and B. In other words.

Mer dist in mid lat = departure.

But by parallel sailing,

Mer dist = D long cos lat.

Hence,

Mer dist in mid lat = D long cos mid lat.

or Dep = D long cos mid lat (I).

But by reference to the triangle for plane sailing,

Dep = dist sin co.

Therefore dist sin co = D long cos mid lat.

or D long = dist sin co sec mid lat (II).

Again, $$\text{Tan co} = \frac{\text{dep.}}{\text{D lat.}}$$

but Dep = D long cos mid lat.

$$\text{Therefore tan co} = \frac{\text{D long cos mid lat}}{\text{D lat.}} \text{ (III).}$$

Great Circle Sailing.

Fig 34.

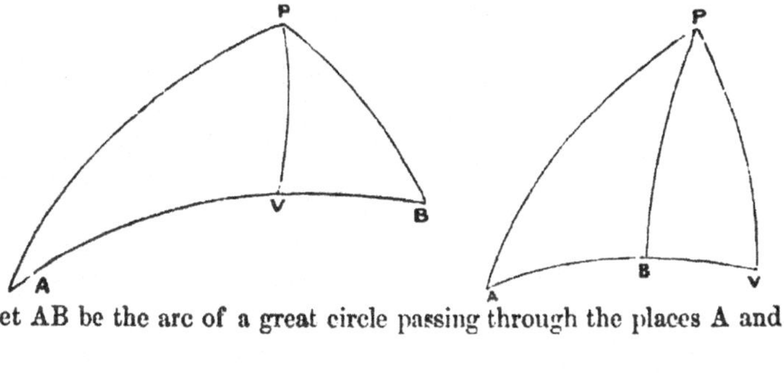

Let AB be the arc of a great circle passing through the places A and B.

Let P be the pole of the earth : then PA and PB are the colatitudes of A and B and the angle APB is the difference of longitude between them. Draw PV perpendicular to AB: then V is the vertex or highest latitude on the great circle track between A and B. In the first figure the point V is on AB ; in the second on AB produced, but in both the principle of the investigation is the same.

Let PB be denoted by a; PA by b; and the angle APB by P; then also AB by d.

$$\text{Then } \cos d = \cos a \cos b + \sin a \sin b \cos P$$

$$\text{but } \cos P = (2 \cos^2 \tfrac{1}{2} P - 1)$$

$$\text{or } \cos d = \cos a \cos b + \sin a \sin b\,(2 \cos^2 \tfrac{1}{2} P - 1)$$

$$\text{,,} \quad = \cos a \cos b - \sin a \sin b + 2 \sin a \sin b \cos^2 \tfrac{1}{2} P$$

$$\text{,,} \quad = \cos (a + b) + 2 \sin a \sin b \cos^2 \tfrac{1}{2} P$$

Subtract both sides from 1, then

$$1 - \cos d = 1 - \cos (a + b) - 2 \sin a \sin b \cos^2 \tfrac{1}{2} P$$

$$\text{or } 2 \sin^2 \frac{d}{2} = 2 \sin^2 \frac{a + b}{2} - 2 \sin a \sin b \cos^2 \tfrac{1}{2} P$$

Let $\sin a \sin b \cos^2 \tfrac{1}{2} P$ be denoted by $\sin^2 \theta$

$$\text{then } \sin^2 \frac{d}{2} = \sin^2 \frac{a + b}{2} - \sin^2 \theta$$

$$\text{whence } \sin \frac{d}{2} = \sqrt{\sin \left\{\frac{a + b}{2} + \theta\right\} \sin \left\{\frac{a + b}{2} - \theta\right\}}$$

which is the expression for the distance.

Again,

$$\frac{\text{Sin AB}}{\text{Sin PB}} = \frac{\sin P}{\sin A}$$

$$\text{or } \sin \text{AB}.\sin A = \sin \text{PB}.\sin P$$

$$\text{and } \sin A = \frac{\sin \text{PB}.\sin P}{\sin \text{AB}}$$

Multiply both sides of the equation by sin PA

then,

$$\text{Sin A}.\sin \text{PA} = \frac{\sin \text{PB}.\sin \text{PA}.\sin P}{\text{Sin AB}}$$

But in the right angled triangle APV, by Napier's rules in Spherical Trigonometry, sin A.sin PA = sin PV

therefore,

$$\text{Sin PV} = \frac{\text{Sin PB}.\sin \text{PA}.\sin P}{\text{Sin AB}}$$

$$\text{or sin co-lat V} = \frac{\text{Sin co-lat B}.\text{sin co-lat A}.\text{sin D.long}}{\text{Sin dist}}$$

whence,

$$\text{cos lat V} = \frac{\text{Cos lat B}.\text{cos lat A}.\text{sin D long}}{\text{Sin dist}}$$

which is the expression for latitude vertex.

Again, in the triangle APV, by Napier's rules,

$$\cos APV = \cot PA . \tan PV$$

$$\text{or } \cos \text{d long from A to V} = \tan \text{lat A} . \cot \text{lat V}$$

by which A is understood to be, in every case, the place of less latitude.

Let x be any point on the great circle between A and the Vertex, and let the arc Px be drawn, then,

$$\text{Cos } VPx = \text{Cot } Px . \tan PV$$

$$\text{or } \cos \text{d long from point to V} = \tan \text{lat } x . \cot \text{lat V}$$

$$\text{whence, } \tan \text{lat } x = \cos \text{d long} . \tan \text{lat V}$$

which is the formula for finding a succession of points on a great circle.

DIP OF THE SEA HORIZON.

Fig. 35.

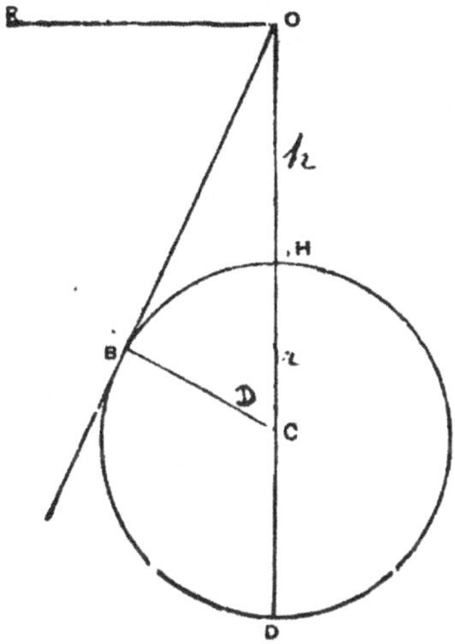

Let O be the position of an observer above the sea level; OC a line passing from the observer's eye through the centre of the earth; OR a horizontal line from the observer; OB a tangent to the earth's surface from O. Join CB. Then by (Euc III—16) the angle OBC will be a right angle, and BOR will, by definition, represent the Dip.

To prove that BCO = the Dip.

Since ROC is a right angle, therefore, ROB and BOC together make a right angle.

Again, since OBC is a right angle, therefore, BOC and BCO together make a right angle. (Euc. I—32.)

Hence, ROB + BOC = BOC + BCO

Omit the common angle BOC

Then ROB = BCO

or BCO is the Dip.

In the investigation, let OH be denoted by h; CB = CH by r, and the angle BCO by D.

$$\text{Then, } \quad \text{Cos D} = \frac{CB}{CO} = \frac{r}{r+h}$$

$$\text{and } 1 - \text{Cos D} = 1 - \frac{r}{r+h} = \frac{r+h-r}{r+h} = \frac{h}{r+h}$$

Now as the radius of the earth r is about 4000 miles, and h is but a few feet, being the height of the observer's eye above the sea, we have,

$$r + h = r \text{ nearly}$$

Therefore,

$$1 - \cos D = \frac{h}{r} \text{ nearly.}$$

$$\text{or } 2 \text{ Sin}^2 \frac{D}{2} = \frac{h}{r} \text{ nearly}$$

Multiply both sides by 2.

$$\text{then } 4 \text{ Sin}^2 \frac{D}{2} = \frac{2h}{r} = \frac{2}{r} \times h$$

Extract the square root of each side.

$$\text{whence } 2 \text{ Sin} \frac{D}{2} = \sqrt{\frac{2}{r}} \times \sqrt{h}$$

Now since it is nearly true that the sine of a small arc taken in minutes is equal to the arc itself multiplied by sin 1′, therefore,

$$\text{Sin} \frac{D'}{2} = \frac{D'}{2} \sin 1', \text{ nearly}$$

$$\text{and } 2 \text{ Sin} \frac{D'}{2} = D' \sin 1'$$

Therefore,

$$D' \text{ Sin } 1' = \sqrt{\frac{2}{r}} \times \sqrt{h}$$

$$\text{whence } D' = \left(\frac{\sqrt{\frac{2}{r}}}{\text{Sin } 1'} \right) \times \sqrt{h}$$

The proper expression for the Dip; the value of the fraction being computed at ·9784 is

$$\text{The Dip in minutes} = \cdot 9784\sqrt{h}$$

For a rough calculation the square root of the height of the eye might be taken as the Dip, thus at 16 feet the Dip, would, in that case, be 4′: but its real value for this number of feet is 3′ 56″, the deficiency of 4″ being caused by the fraction $\left(\frac{\sqrt{\frac{2}{r}}}{\text{Sin } 1'} \right)$ or ·9784.

From this comes the following useful little deduction, by which the distance of the visible horizon at sea may be found. Supposing, as before, that CH $= r$, and OH $= h$, then, by (Euc III — 36).

$$DO.OH = OB^2 \text{ (or } d^2 \text{)}$$

$$\text{or } (2r + h).h = d^2$$

$$\text{and } 2rh + h^2 = d^2$$

Omit h^2 it being small

$$\text{then } 2rh = d^2$$

Now r being 4000 miles nearly, $2r$ is 8000 miles; and h represented as the fraction of a foot is $\frac{h}{5280}$, thence,

$$8000 \times \frac{h}{5280} = d^2$$

$$\text{this being reduced} = \frac{3h}{2} \text{ nearly} = h + \tfrac{1}{2} h$$

$$\text{therefore } d = \sqrt{h + \tfrac{1}{2} h}$$

Or, practically, the distance of the visible horizon at sea may be found by adding to the height of the eye its half, and taking the square root of the sum: thus, at an elevation of 24 feet we have $\sqrt{24 + 12} = \sqrt{36} = 6$ miles nearly.

Parallax in Altitude.

Fig. 36.

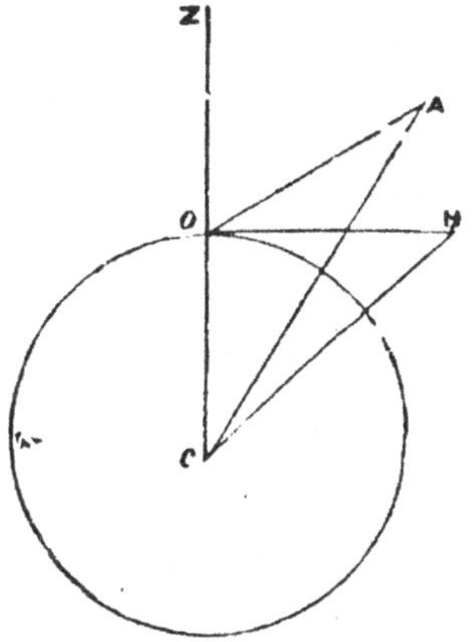

Let O be the place of an observer, A the position of a heavenly body in altitude, and H its place on the sensible horizon, then the angle OAc is called its "parallax in altitude," and OHc its "horizontal parallax," and AOH is the apparent altitude of A.

Then in the right angled triangle OcH,

$$\frac{CO}{CH} = \sin H = \sin \text{Hor}^{l} \text{Par}^{x}$$

$$\frac{CO}{CA} = \frac{\sin A}{\sin COA} = \frac{\sin A}{\sin ZOA} = \frac{\sin A}{\cos AOH}$$

COA and ZOA being supplements, and ZOA and AOH being complements

$$\text{But } \frac{CO}{CH} = \frac{CO}{CA}$$

$$\text{Therefore, } \sin \text{Hor}^{l} \text{Par}^{x} = \frac{\text{Sin Par}^{x} \text{ in alt}}{\text{Cos app}^{t} \text{ alt}}$$

$$\text{Whence, } \sin \text{Par}^{x} \text{ in alt} = \sin \text{Hor}^{l} \text{Par}^{x} \times \cos \text{app}^{t} \text{ alt.}$$

Making use of the preceding value for the sine of a small arc, but, this time, measuring the arc in seconds, we have,

$$\text{Sin Par}^{x} \text{ in alt}'' = \text{Par}^{x} \text{ in alt}'' \times \text{Sin } 1''$$

$$\text{and Sin Hor}^{l} \text{Par}^{x''} = \text{Hor}^{l} \text{Par}^{x''} \times \text{Sin } 1''$$

whence

$$\text{Par}^x \text{ in alt}'' \text{ Sin } 1'' = \text{Hor}^l \text{ Par}^{x''} \text{ Sin } 1'' \text{ Cos app}^t \text{ alt}$$
$$\text{or Par}^l \text{ in alt}'' = \text{Hor}^l \text{ Par}^{x''} \text{ Cos app}^t \text{ alt}$$

which though but an approximation is sufficiently near for all practical purposes.

Here it is evident, as the figure shows, that the Horizontal Parallax is the greatest parallax, for since the cosine diminishes a quantity, (being a proper fraction,) when multiplied into it we have Horizontal parallax greater than parallax in altitude, the former being diminished by cos alt to equal the latter.

Latitude by the Meridian Altitude of a Body above and below Pole.

Fig. 37.

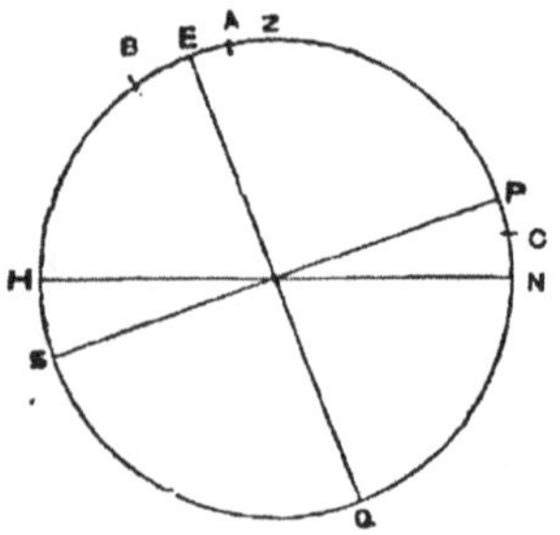

Let the circle represent the meridian of the observer, Z the zenith, HN his horizon, EQ the celestial equator, P, S, the elevated and depressed poles of the celestial sphere, A, B, and C, positions of a heavenly body above and below pole.

(1) To show that the elevation of the pole above the horizon is the latitude.

or PN = lat.

Since, on the earth's surface, the arc of a meridian between the observer's position and the equator is the latitude of the place, therefore, the corresponding arc of the celestial meridian between his zenith and the celestial equator is also the latitude, therefore

ZE = lat; then ZN = EP, each being a quadrant,
or EZ + ZP = ZP + PN
omit ZP,
then EZ = PN
or PN = lat of place.

(2) Let the celestial object be upon the meridian at A, when its zenith distance and declination are of the same name. For, supposing P to be the North Pole, the zenith Z would be north of the object A, and the declination EA, always reckoned from the equator, would also be north, then,

ZE = ZA + AE
or lat = zen dist + declination.

(3). Let the celestial object be on the meridian at B, when its Zenith distance and declination are of contrary names. For supposing P to be the North pole, the Zenith would then be North of the body at A, but the declination EB, always reckoned from the equator, would now be South; then,

$$ZE = ZB - BE$$

$$\text{or lat} = \text{zen dist} - \text{declination}$$

(4). Let the celestial object be on the meridian below pole as at C, then PC is its co-declination (polar distance), and CN is its altitude under the pole, therefore,

$$PN = PC + CN$$

$$\text{or lat} = \text{polar dist} + \text{altitude}$$

From this it is evident that, for an object to be visible under the pole, its polar distance PC, must be less than the latitude of the place of observation, in which case it describes a circle around the pole and is termed circumpolar; thus in June when the Sun's declination is 23° 28′ its polar distance is 66° 32′, and it will be visible under the pole in any latitude greater than that amount.

Latitude by an Observation of the Pole Star out of the Meridian.

Fig. 38.

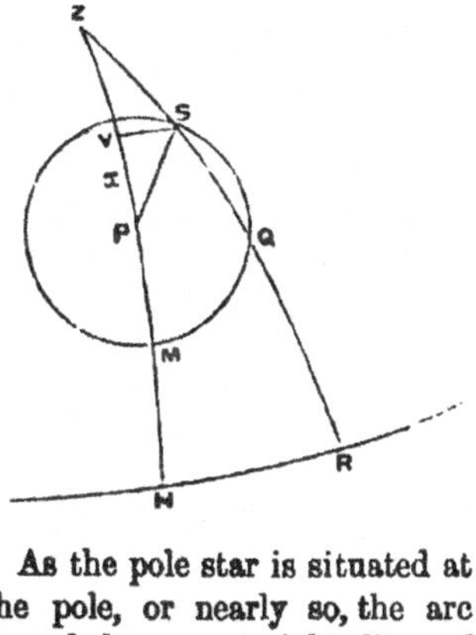

Let Z be the Zenith; P the pole; SQM the diurnal circle described by the pole star: S the place of the star when observed; ZSR a circle of altitude over its place; join PS, and draw SV perpendicular to ZH.

Then since the arc ZS is greater than ZV, a circle described with centre Z and radius ZS, will cut below V: suppose it to cut in x, then PV is the first correction, and Vx the second.

As the pole star is situated at the distance of a degree and a half from the pole, or nearly so, the arc forming its polar distance PS may be regarded as a straight line, when compared with the curvature of the sphere, and the triangle VPS may be assumed to be a plane triangle, whence

$$VP = PS . \cos VPS$$

$$\text{or 1st cor} = \text{star's polar dist} \times \cos \text{hour angle}.$$

$$\text{Again } ZS = Zx = (ZV + Vx)$$

$$\text{and } \cos ZS = \cos ZV \cos Vx - \sin ZV \sin Vx$$

$$\text{omit } \cos Vx,\ Vx \text{ being small,}$$

$$\text{then } \cos ZS = \cos ZV - \sin ZV . \sin Vx$$

but, in the right angled spherical triangle ZVS,

$$\cos ZS = \cos ZV . \cos VS$$

$$\text{therefore } \cos ZV \cos VS = \cos ZV - \sin ZV \sin Vx$$

$$\text{or } \sin ZV \sin Vx = \cos ZV - \cos ZV \cos VS$$

$$\text{,,} \qquad = \cos ZV . (1 - \cos VS)$$

$$\text{,,} \qquad = \cos ZV . 2 \sin^2 \frac{VS}{2}$$

$$\text{and } \sin Vx = \cot ZV . 2 \sin^2 \frac{VS}{2}$$

$$\text{But } \operatorname{Sin} Vx'' = Vx'' \operatorname{Sin} 1'', \ (Vx \text{ being small})$$

$$\text{,, } \operatorname{Cot} ZV = \tan VH = \tan \text{alt nearly}$$

$$\text{and } \operatorname{Sin} \frac{VS''}{2} = \frac{VS''}{2} \operatorname{Sin} 1''$$

$$\text{or } \operatorname{Sin}^2 \frac{VS}{2} = \frac{VS^2}{4} \operatorname{Sin}^2 1''$$

$$\text{,, } 2 \operatorname{Sin}^2 \frac{VS}{2} = \frac{VS^2}{2} \operatorname{Sin}^2 1'' = \frac{(p \operatorname{Sin} h)^2}{2} \operatorname{Sin}^2 1''$$

Hence,

$$Vx'' \operatorname{Sin} 1'' = \tan \text{alt} \frac{(p \operatorname{Sin} h)^2}{2} \operatorname{Sin}^2 1''$$

$$\text{or 2nd Cor}'' = \tfrac{1}{2} \operatorname{Sin} 1'' \tan \text{alt} \, (p \operatorname{Sin} h)^2$$

By reference to the preceding figure it will be seen, that, when the arc PS is at right angles to the meridian, or the hour angle is either 6 hours or 18 hours, then S is on a level with P and the altitude of the star is the latitude of the place. But should the star be anywhere on the upper half of the circle, the altitude of the star is too great for that of the pole, or the latitude: in such cases the first correction is negative; on the other hand, if the star be in the lower half of the circle, its altitude is too small for the latitude, and the correction is added to it; or practically, when the hour angle is from 6 to 18 hours the first correction is added: between 0^h and 6^h subtracted, also between 18^h and 0^h.

Latitude by Reduction.

Fig. 39.

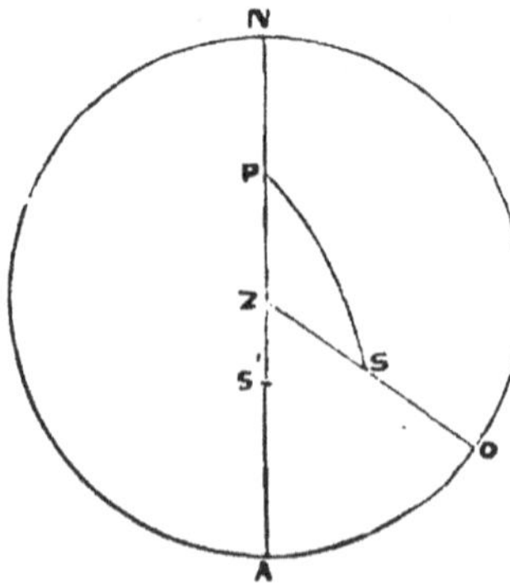

Let NPZ represent the meridian, S the place of an object near it at the time of observation: then, PN = lat: PZ = co-lat: PS = object's polar distance; ZS = zenith distance; ZS' its meridian zenith distance, S' being its place on the meridian. Suppose the polar distance PS not to change between S and S'.

Let PZ $= l'$: PS $= p$: ZS $= z$: ZS$' = z'$: ZPS $= h$.
then by supposition,

$$PS = PS'$$
$$\text{or } p = z' + l'$$
$$\text{and } z' = p - l'$$
$$\cos z' = \cos p \cos l' + \sin p \sin l'$$

In the triangle ZPS, $\cos z = \cos p \cos l' + \sin p \sin l' \cos h$

$$\text{then } \cos z' - \cos z = \sin p \sin l' (1 - \cos h)$$

$$\text{or } 2 \sin \frac{z + z'}{2} \sin \frac{z - z'}{2} = \sin p \sin l' \, 2 \sin^2 \tfrac{1}{2} h$$

$$\text{But since } z = z' \text{ nearly } \frac{z + z'}{2} = z' \text{ nearly}$$

$$\text{and} \qquad \frac{z - z'}{2} = \frac{\text{red}^n}{2}$$

For the reduction is the difference between the meridian zenith distance and that at the time of observation, therefore,

$$2 \sin z' . \sin \frac{r}{2} = \cos \text{dec} \cos \text{lat} \,. 2 \sin^2 \tfrac{1}{2} h$$

$$\text{Now } 2 \sin \frac{r''}{2} = r'' \sin 1''$$

$$\text{whence } r'' = \cos \text{dec} \cos \text{lat} \operatorname{cosec} \text{MZD}. \frac{2 \sin^2 \frac{1}{2} h}{\sin 1''}$$

The Sun's meridian altitude is always his greatest altitude, and therefore the reduction is additive to the ex-meridian altitude, or subtractive from the zenith distance. The moon's meridian altitude is only her greatest when she is in extreme declination in the hemisphere of the observer. Reduction by the moon is not a method to be recommended.

Hour Angles, Meridian Distances, &c.

Fig. 40.

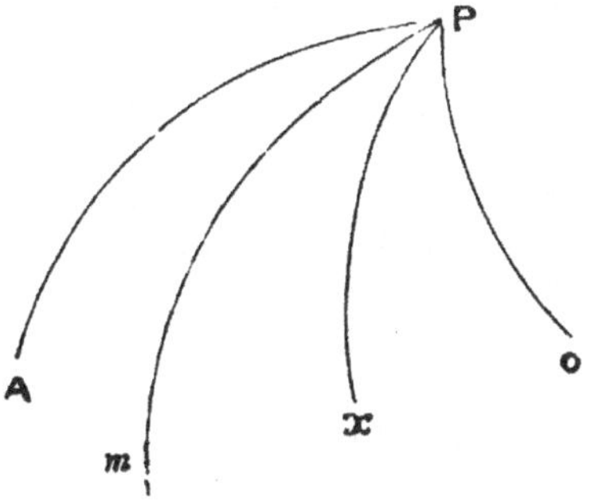

Let A be the meridian of the first point of Aries, m that of the mean Sun, x of any object, and PO that of the observer, then,

$$APO = APm + mPO$$

(1). Or R*a* of meridian = R*a* of Mean Sun + Mean Time.

$$\text{Again } xPo = APo - APx$$

(2) Westerly hour angle of x = RA of mer — RA of x.

Also $xPo + APx = APo$

(3) Westerly meridian distance of x + Ra of x = Ra of meridian.

Investigation for Hour Angle.

In figure for reduction to the meridian, and the triangle ZPS, we have,

$$\text{Cos } h = \frac{\text{Cos } z - \cos l' \cos p}{\sin l' \sin p}$$

$$\text{or Cos } h = \frac{\text{Sin alt} - \text{sin lat} \cos p}{\text{Cos lat} \sin p}$$

Subtract both sides from 1, then,

$$1 - \cos h = 1 - \frac{\text{Sin } a - \sin l \cos p}{\cos l \sin \text{p}}$$

$$\text{,,} \quad = \frac{\text{Cos } l \sin p - \sin a + \sin l \cos p}{\cos l \sin p}$$

$$\text{,,} \quad = \frac{\text{Sin } (p + l) - \sin a}{\cos l \sin p}$$

$$\text{,,} \quad = \frac{2 \cos \frac{p + l + a}{2} . \sin \frac{p + l - a}{2}}{\cos l \sin p}$$

Now if $\frac{p + l + a}{2}$ be represented by S

then $\frac{p + l - a}{2}$ is (S — a)

and $1 - \cos h$ is $2 \sin^2 \frac{1}{2} h$

$$\text{Hence } \sin \frac{h}{2} = \sqrt{\sec l \operatorname{cosec} p . \cos \text{S} . \sin (\text{S} - a)}$$

The Westerly hour angle of the apparent or true sun measures apparent time; to turn this into mean time the equation of time must be applied. The Westerly hour angle of any other body + the body's Ra — mean sun's Ra = the mean time at place; (see preceding figure).

To Compute an Objects Azimuth.

In figure for Reduction the angle PZS or AZS is the Azimuth, the distinction being, that, the former is named with the latitude, and the latter contrary to it. Then if PZ = l', and ZS = z as before, we have,

$$\text{Cos PZS} = \frac{\cos p - \cos l' \cos z}{\text{Sin } l' \sin z}$$

but Cos PZS = — cos AZS

$$\text{therefore} - \text{Cos AZS} = \frac{\cos p - \sin l \sin a}{\text{Cos } l \cos a}$$

add 1 to both sides, then,

$$1 - \cos AZS = 1 + \frac{\cos p - \sin l \sin a}{\text{Cos } l \cos a}$$

$$,, \quad = \frac{\text{Cos } l \cos a - \sin l \sin a + \cos p}{\text{Cos } l \cos a}$$

$$,, \quad = \frac{\text{Cos } (a + l) + \cos p}{\text{Cos } l \cos a}$$

$$,, \quad \text{or } 2 \text{ Sin}^2 \frac{AZS}{2} = \frac{2 \text{ Cos} \frac{a + l + p}{2} \cos \frac{a + l - p}{2}}{\text{Cos } l \cos a}$$

$$\text{whence Sin} \frac{AZS}{2} = \sqrt{\text{Sec } l \sec a \cos S . \cos (S - p)}$$

To Compute the Amplitude of a body.

Fig. 41.

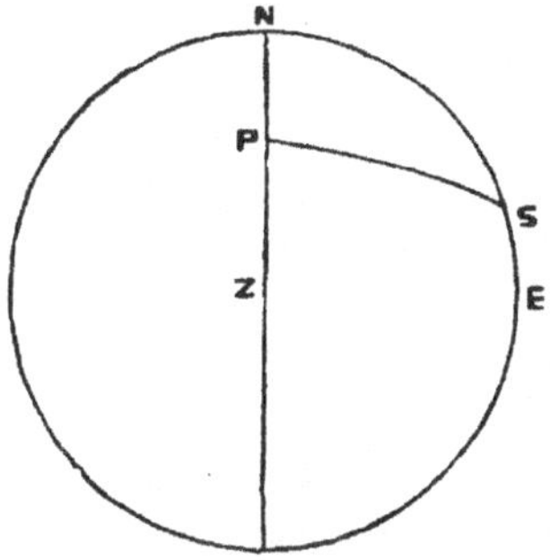

Let NZ be the meridian of the observer, S the place of a body rising or setting, E the east or west points of the horizon, then ES is its amplitude, and by Napier's Rules we have in the triangle PNS,—

$$\text{Cos PS} = \cos \text{PN} \cos \text{NS}$$
$$\text{Cos } p = \cos \text{lat} \sin \text{SE}$$
$$\text{whence Sin dec} = \cos \text{lat} \sin \text{amp}$$
$$\text{and Sin amp} = \sin \text{dec} \sec \text{lat}$$

Ivory's Method of Double Altitudes.

Fig. 42

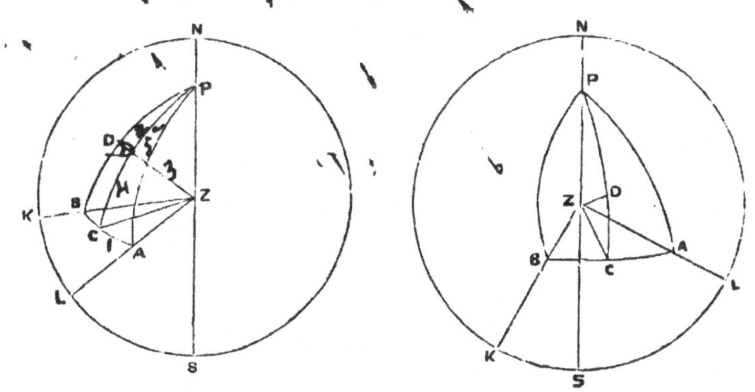

In the first of these figures, the object is supposed to be on the same side of the meridian at both observations: in the second, on opposite sides when observed. Let the circles represent the horizon, NS the meridian of the observer, P the pole, Z the zenith, B and A the places of the object when observed, then PA and PB, their polar distances, are supposed to be equal; ZK and ZL two vertical circles passing over A and B. Bisect the angle BPA by PC, which, if PA and PB are equal, will bisect BA in C; join CZ, and draw ZD perpendicular to PC.

Then AC is arc 1: CP arc 2: ZD arc 3: CD arc 4: DP arc 5: and PZ is the co-latitude. If the object observed be the Sun, which is generally the case, the angle BPA is the interval of apparent time between the observations, and APC is half the elapsed time which may be denoted by H. Let the polar distances PA and PB be denoted by p, and the altitudes AL and BK by a and b.

For Arc 1

In the right angled spherical triangle ACP.

$$\text{Sin AC} = \sin \text{AP} . \sin \text{APC}.$$
$$\text{or Sin arc } 1 = \sin p . \sin \text{H}$$

For Arc 2

$$\text{Cos AP} = \cos \text{AC} . \cos \text{CP}$$
$$\text{or Cos } p = \cos \text{arc } 1 . \cos \text{arc } 2.$$
$$\text{and Cos } p . \sec \text{arc } 1 = \cos \text{arc } 2$$

For Arcs 3 and 4

In the triangles ZCA and ZCB,

$$\text{Cos ZA} = \cos \text{ZC} . \cos \text{CA} + \sin \text{ZC} . \sin \text{CA} . \cos \text{ZCA}$$
$$\text{and Cos ZB} = \cos \text{ZC} . \cos \text{CB} + \sin \text{ZC} . \sin \text{CB} . \cos \text{ZCB}$$
$$\text{But } \cos \text{ZA} = \sin \text{AL} = \sin a$$
$$\text{and } \cos \text{ZB} = \sin \text{BK} = \sin b$$
also CB and CA are equal
$$\text{and } \cos \text{ZCA} = \sin \text{ZCD (being complements)}$$
$$\text{,, } \cos \text{ZCB} = - \cos \text{ZCA} = - \sin \text{ZCD}.$$

Therefore, substituting,

$$\text{Sin } a = \cos \text{ZC} . \cos \text{CA} + \sin \text{ZC} . \sin \text{CA} . \sin \text{ZCD}$$
$$\text{Sin } b = \cos \text{ZC} . \cos \text{CA} - \sin \text{ZC} . \sin \text{CA} . \sin \text{ZCD}$$

Subtracting one equation from the other,

$$\text{Sin } a - \sin b = 2 \sin \text{ZC} . \sin \text{CA} . \sin \text{ZCD}$$
$$\text{But } \sin \text{ZC} . \sin \text{ZCD} = \sin \text{ZD}$$
for ZDC is a right angle.
$$\text{and } \sin a - \sin b = 2 \cos \frac{a+b}{2} . \sin \frac{a-b}{2}$$
$$\text{then, } 2 \cos \frac{a+b}{2} . \sin \frac{a-b}{2} = 2 \sin \text{ZD} . \sin \text{CA}$$

Let half the sum of the altitudes be denoted by S, and half their difference *by* D; then,

$$\text{Cos S} . \sin \text{D} = \sin \text{arc } 3 . \sin \text{arc } 1.$$
$$\text{Whence, } \cos \text{S} . \sin \text{D} \operatorname{cosec} \text{arc } 1 = \sin \text{arc } 3.$$

Again, adding the equations together, we have

$$\text{Sin } a + \sin b = 2 \cos ZC . \cos CA.$$

$$\text{But } \cos ZC = \cos ZD . \cos DC$$

$$\text{and Sin } a + \sin b = 2 \sin \frac{a+b}{2} . \cos \frac{a-b}{2}$$

$$\text{then } 2 \sin \frac{a+b}{2} \cos \frac{a-b}{2} = 2 \cos ZD . \cos DC . \cos CA$$

or sin S.cos D = cos arc [illegible].cos arc [illegible].cos arc [illegible]

and sin S cos D.sec arc 3.sec arc 1 = cos arc 4.

For Arc 5.

CP = CD + DP

or arc [illegible] = arc [illegible] + arc 5

and arc [illegible] — arc [illegible] = arc 5

Latitude.

In the right angled spherical triangle PDZ,

cos PZ = cos PD.cos DZ

or sin lat = cos arc 5.cos arc [illegible].

SUMNER'S METHOD OF DOUBLE ALTITUDES.

Fig. 43.

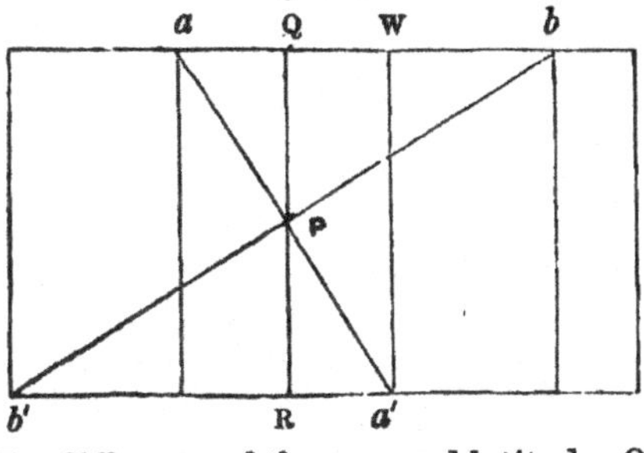

Let the lines of position be represented by *aa'* and *bb'*, then P, their intersection, will be the place of the observer.

Let the d long from *a* to *a'*, viz. *a*W, be denoted by C.

The difference of the assumed latitudes QR by D; *ab* by B, and *a' b'* by A,

Then, in the similar triangles *a*P*b* and *a'*P*b'*,

$$\frac{ab}{a'b'} = \frac{PR}{PQ}$$

adding 1 to both sides,

$$\frac{a'b' + ab}{a'b'} = \frac{PR + PQ}{PQ} = \frac{QR}{PQ} = \frac{D}{PQ}$$

$$\text{ie } \frac{A + B}{A} = \frac{D}{PQ} = \frac{D}{\text{Cor for lat}}$$

$$\text{whence, Cor for lat} = \frac{AD}{A + B}$$

Again, in the similar triangles *a*QP and *a*W*a'*,

$$\frac{aQ}{QP}=\frac{aW}{Wa'}$$

$$\text{whence } aQ=\frac{aW \times QP}{a'W}=\frac{C \times QP}{QR}=\frac{C \times QP}{D}$$

$$\text{or Cor for long}=\frac{C \times \frac{A\,D}{A+B}}{D}=\frac{A\,C}{A+B}$$

To clear a Lunar Distance from the effects of Parallax and Refraction.

Fig. 44.

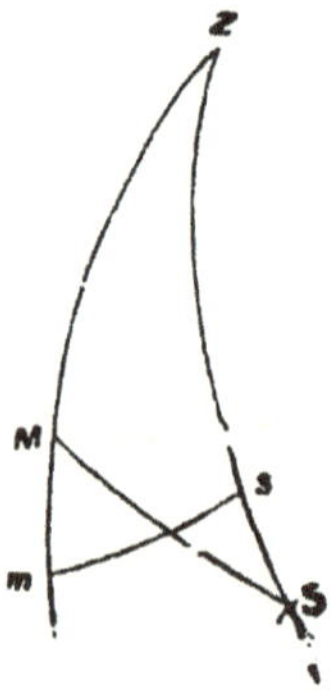

Let Z be the Zenith, ZMm and ZsS two vertical circles passing over the positions of the objects.

I. In the figure, let

M represent the true place of the Moon
m „ „ app. „ „ „
S „ „ true „ „ Sun
s „ „ app. „ „ „
or of a Star.

Then, since the Moon's parallax is greater than her refraction, M is above m. For an opposite reason, s is above S.

II. In the following investigation, let

M° = the true altitude of the Moon,
m° = „ app. „ „ „
S° = the true altitude of the Sun,
s° = „ app. „ „ „

then, $ms = d$, the apparent distance, and MS = D the true distance. In the triangle Zms.

$$\text{Cos}\frac{Z}{2}=\sqrt{\frac{\text{Sin}\ \frac{1}{2}\,(Zm+Zs+ms)\ \sin\ \frac{1}{2}\,(Zm+Zs-ms)}{\text{Sin}\ Zm\ \sin\ Zs}}$$

$$\text{or Cos}^2\frac{Z}{2}=\text{Sin}\,\frac{Zm+Zs+ms}{2}\ \sin\,\frac{Zm+Zs-ms}{2}\ \text{cosec}\ Zm\ \text{cosec}\ Zs$$

$$„\ =\text{Sin}\frac{\overline{90-m}+\overline{90-s}+d}{2}.\sin\frac{\overline{90-m}+\overline{90-s}-d}{2}.\text{cosec}\ \overline{90-m}.\ \text{cosec}\ \overline{90-s}.$$

$$„\ =\text{Sin}\frac{180-\overline{m+s-d}}{2}.\sin\frac{180-\overline{m+s+d}}{2}.\sec m.\ \sec s$$

$$„\ =\text{Sin}\left\{90-\frac{m+s-d}{2}\right\}.\sin\left\{90-\frac{m+s+d}{2}\right\}.\sec m\ \sec s$$

$$„\ =\text{Cos}\ \frac{m+s-d}{2}.\cos\frac{m+s+d}{2}.\sec m\ \sec s$$

$$\text{Let } \frac{m+s+d}{2} \text{ be denoted by X}$$

$$\text{then } \frac{m+s-d}{2} \text{ is } (X-d)$$

therefore,

$$\text{Cos}^2\frac{Z}{2} = \cos X . \cos (X-d) . \sec m . \sec s.$$

In the triangle ZMS, let

$$\text{Sin}^2 \theta = \sin ZM . \sin ZS \cos^2 \tfrac{1}{2} Z$$

$$\text{then } \text{Sin}^2\frac{MS}{2} = \sin\left\{\frac{ZM + ZS}{2} + \theta\right\} . \sin\left\{\frac{ZM + ZS}{2} - \theta\right\}$$

$$\text{hence, } \text{Sin}^2 \theta = \cos M . \cos S . \cos X . \cos (X-d) \sec m \sec s.$$

$$\text{and } \text{Sin}^2\frac{D}{2} = \sin\left\{\frac{90-M+90-S}{2}+\theta\right\} \sin\left\{\frac{90-M+90-S}{2}-\theta\right\}$$

$$\text{,,} \quad = \text{Sin}\left\{\frac{180-\overline{M+S}}{2}+\theta\right\} \sin\left\{\frac{180-\overline{M+S}}{2}-\theta\right\}$$

$$\text{,,} \quad = \text{Sin}\left\{90-\left(\frac{M+S}{2}-\theta\right)\right\} . \sin\left\{90-\left(\frac{M+S}{2}+\theta\right)\right\}$$

$$\text{,,} \quad = \text{Cos}\left\{\frac{M+S}{2}-\theta\right\} . \cos\left\{\frac{M+S}{2}+\theta\right\}$$

The Lunar distance might also be cleared by the direct application of the rules of spherical Trigonometry.

Thus,

In the triangle Zms, Zm, Zs and ms being known, angle Z might be found, for,

$$\text{Cos}\frac{Z}{2} = \sqrt{\frac{\text{Sin}\frac{Zm + Zs + ms}{2} . \sin\frac{Zm + Zs - ms}{2}}{\text{Sin } Zm \sin Zs}}$$

Then in the triangle ZMS, Z being known, and also ZM, ZS, the side MS might be found, for,

$$\text{Sin}^2 \theta = \sin ZM . \sin ZS . \cos^2 \tfrac{1}{2}Z$$

$$\text{and } \text{Sin}^2\frac{MS}{2} = \sin\left\{\frac{ZM + ZS}{2} + \theta\right\} \sin\left\{\frac{ZM + ZS}{2} - \theta\right\}$$

It will be seen, that, the method as above investigated, is a compound of these two formulæ of spherical Trigonometry.

Equation of Equal Altitudes.

The equation of equal altitudes is a correction to be applied to the middle time by a chronometer obtained from the times it showed when the Sun had equal altitudes east and west of the meridian. On account of the change which takes place in the Sun's polar distance in the interval between two equal altitudes of that body, the hour angles at the times of observation are not equal. Half the difference of these hour angles is the equation of equal altitudes.

Resuming the equation obtained from the triangle ZPS, (see figure for Reduction to the Meridian), we have,

$$\text{Cos } z = \cos p \cos l' + \sin p \sin l' . \cos h$$

Now if the polar distance be **increased** by any quantity as c, the hour angle will be **decreased** by a corresponding quantity e, which is the correction sought.

Let p become then, $(p + c)$
and h will be $(h - e)$

Therefore.

$$\text{Cos } z = \cos (p + c) \cos l' + \sin (p + c) \sin l' . \cos (h - e)$$

Expanding, we have

$$\text{Cos} z = [\cos p \cos c - \sin p \sin c] . \cos l' + [\sin p \cos c + \cos p \sin c] . \sin l' . \cos (h - e).$$

Then,

$\text{Cos } z = \cos p \cos c \cos l' - \sin p \sin c \cos l' + [\sin p \cos c \sin l' + \cos p \sin c \sin l'] \cos (h - e)$

$\cos c$ and $\cos e$ may be omitted, each being equal to 1 nearly.

$= \cos p \cos l' - \sin p \sin c \cos l + [\sin p \sin l' + \cos p \sin c \sin l'] [\cos h + \sin h \sin e]$

$= \cos p \cos l - \sin p \sin c \cos l' + \sin p \sin l' \cos h + \sin p \sin l \sin h \sin e + \cos p \sin c \sin l' \cos h + \cos p \sin c \sin l' \sin h \sin e$

The latter term may be omitted as containing the sines of c and e, both small arcs,

And $\cos z = \cos p \cos l' + \sin p \sin l' \cos h$, these equivalents cancel each other, then

$$O = - \sin p \sin c \cos l' + \sin p \sin l' \sin h \sin e + \cos p \sin c \sin l' \cos h$$

Transposing the term containing $\sin e$, and allowing for change of sign, we have,

$$\text{Sin } p \sin l' \sin h \sin e = \sin c [\sin p \cos l' - \cos p \sin l' \cos h]$$

$$\text{or Sin } e = \sin c \left\{ \frac{\sin p \cos l'}{\sin p \sin l' \sin h} - \frac{\cos p \sin l' \cos h}{\sin p \sin l' \sin h} \right\}$$

$$= \sin c [\cot l' \text{cosec } h - \cot p \cot h]$$

Whence

$$e = c \tan l \text{ cosec } h - c \cot p \cot h$$

These parts in the practical work are A and B, and are added or subtracted thus; — if p be less than 90° as assumed in the above, then cot p is + and the parts are subtracted; but, if p be obtuse, cot p is —, and the parts are additive, thus

A = c tan lat cosec h, and
B = c tan dec cot h,
A + B when p is greater than 90°,
A — B „ „ less „

Or otherwise—

A + B when the lat and dec are of opposite names
A — B „ „ „ the same name.

THE EFFECT PRODUCED UPON THE HOUR ANGLE, BY AN ERROR IN THE ALTITUDE FROM WHICH IT IS OBTAINED.

Resuming the equation,

$$\text{Cos } z = \cos p \cos l' + \sin p \sin l' \cos h$$

Let the altitude be decreased, or the zenith dsstance increased, by any quantity x, then h the hour angle is increased by a corresponding amount y, then,

$$\text{Cos } (z + x) = \cos p \cos l' + \sin p \sin l' \cos (h + y)$$

or $\cos z \cos x - \sin z \sin x = \cos p \cos l' + \sin p \sin l' [\cos h \cos y - \sin h \sin y]$

Eliminate $\cos x$ and $\cos y$ the increments,

therefore,

$$\cos z - \sin z \sin x = \cos p \cos l' + \sin p \sin l' \cos h - \sin p \sin l' \sin h \sin y$$

but $\cos z = \cos p \cos l' + \sin p \sin l' \cos h$

hence $\sin z \sin x = \sin p \sin l' \sin h \sin y$

and $$\frac{\sin z}{\sin p \sin l' \sin h} = \frac{\sin y}{\sin x}$$

But by reference to the figure, it will be seen that,

$$\sin p : \sin Z :: \sin z : \sin h$$

$$\text{or } \sin p \sin h = \sin z \sin Z$$

therefore, $$\frac{\sin z}{\sin l' \sin z \sin Z} = \frac{y}{x} \text{ nearly}$$

or $$\frac{1}{\cos \text{lat} \sin Az} = \frac{\text{Error in Hour Angle}}{\text{Error in altitude}}$$

whence, Error in Hour Angle = error in alt sec lat cosec Az

This becomes a minimum when the body is on or near the prime vertical, and a maximum when on the meridian.

COMPUTATION OF ALTITUDES.

First Method.

$$\begin{aligned}\text{Cos } z &= \cos p \cos l' + \sin p \sin l' \cos h \\ &= \cos p \cos l' + \sin p \sin l' .(2 \cos^2 \tfrac{1}{2}h - 1) \\ &= \cos p \cos l' - \sin p \sin l' + \sin p \sin l' . 2 \cos^2 \tfrac{1}{2}h \\ &= \cos (p + l') + 2 \sin p \sin l' \cos^2 \tfrac{1}{2}h\end{aligned}$$

Let $\sin p \sin l' \cos^2 \frac{1}{2}h = \sin^2 \theta$

then $\cos z = \cos (p + l') + 2 \sin^2 \theta$

Subtract both sides from 1, then,

$$1 - \cos z = 1 - \cos (p + l') - 2 \sin^2 \theta$$

$$\text{or } 2 \sin^2 \frac{z}{2} = 2 \sin^2 \frac{p + l'}{2} - 2 \sin^2 \theta$$

$$\text{and Sin } \frac{z}{2} = \sqrt{\text{Sin}\left\{\frac{p + l'}{2} + \theta\right\} \sin\left\{\frac{p + l'}{2} - \theta\right\}}$$

Second Method.

Fig. 45.

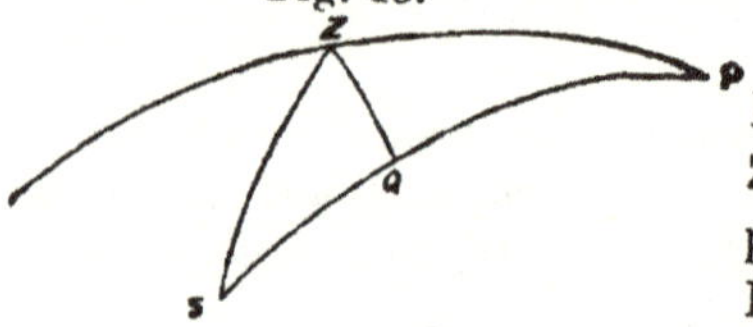

If PZ represents the meridian, PS the polar distance of S, and ZS its zenith distance, and a perpendicular be drawn from Z on PS, viz ZQ, we have,

$$\text{Cos } h = \cot PZ \,.\, \tan PQ$$

$$\text{or Cos } h \,.\, \cot \text{lat} = \tan PQ \text{ (call this arc 1)}$$

$$\text{and } PS - PQ = QS : \textit{ie } p - \text{arc } 1 = QS \text{ (arc 2)}$$

$$\text{then Cos } PZ : \cos ZS :: \cos PQ : \cos QS$$

$$\text{therefore, Sin lat} : \sin \text{alt} :: \cos \text{arc } 1 : \cos \text{arc } 2.$$

$$\text{whence, Sin alt} = \cos \text{arc } 2 \,.\, \sin \text{lat} \,.\, \sec \text{arc } 1.$$

Time of Sunrise.

Fig. 46.

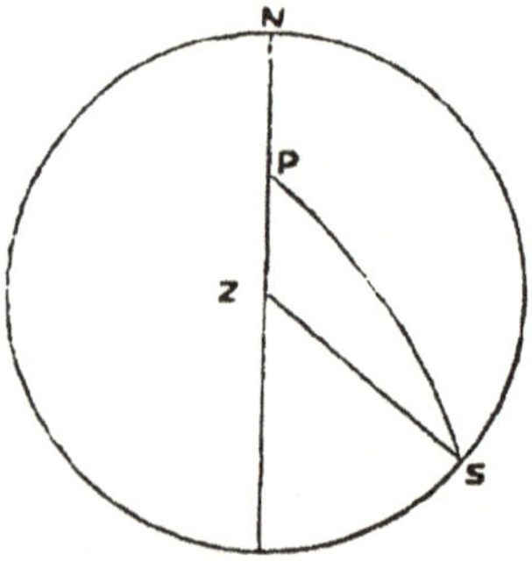

Let S be the place of the Sun when on the horizon,* then PS is its polar distance, ZS its zenith distance, and PZ the co-latitude, and ZS = 90°; ZPS the hour angle at Sunset = h.

Then,

$$\text{Cos } h = \frac{\cos ZS - \cos ZP \cos PS}{\text{Sin } ZP \,.\, \sin PS}$$

$$= \frac{\text{Cos } 90° - \cos l' \,.\, \cos p}{\text{Sin } l' \sin p} = -\cot l' \cot p$$

$$\text{or Cos } h = -\tan \text{lat} \cot p$$

The negative sign is thus accounted for ;—when p is obtuse, cot p is negative, and then cos h becomes positive ; when p is less than 90°, cos h is negative and h is subtracted from 180°.

Time of a Star's Rising or Setting.

Resuming the Equation (See Hour Angle), we have

$$2 \text{ Sin}^2 \frac{h}{2} = \frac{\text{Sin } (p + l) - \sin a}{\text{Cos } l \sin p}$$

The dip and refraction will cause the star to appear on the horizon when it is really depressed below it : call this depression, or negative altitude, d, then the step becomes,

$$2 \text{ Sin}^2 \frac{h}{2} = \frac{\text{Sin } (p + l) + \sin d}{\text{Cos } l \sin p}$$

Whence,

$$\text{Sin } \tfrac{1}{2} h = \sqrt{\text{Sin S} \cos (s - d) \sec l \,\text{cosec}\, p}$$

The Position of a Body upon Special Circles of the Heavens.

Fig. 47.

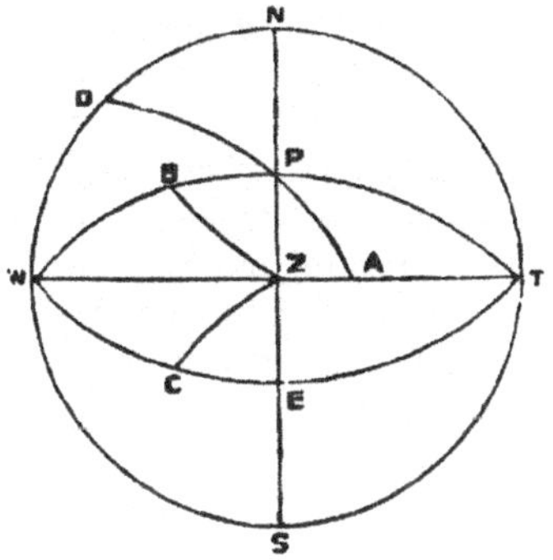

Let NTSW represent the horizon; NS the celestial meridian, WZT the prime vertical, WPT the six o'clock hour circle, WET the equator.

A the place of a heavenly body on the Prime Vertical, B its position on the six o'clock hour circle, C on the equator and D on the horizon, then,

(1). When at A, on the Prime Vertical.

$$\text{Cos PA} = \cos \text{PZ} \cos \text{ZA}$$
$$\text{or Cos } p = \cos l \cos z$$
$$\text{and Sin dec} = \sin \text{lat} \sin \text{alt } (a)$$
$$\text{also Cos APZ} = \cot \text{PA} \tan \text{PZ}$$
$$\text{or Cos } h = \tan \text{dec} \cot \text{lat } (b)$$

(2). When at B, on the six o'clock hour circle.

$$\text{Cos ZB} = \cos \text{ZP} \cos \text{PB}$$
$$\text{or Sin alt} = \sin \text{lat} \sin \text{dec } (c)$$
$$\text{and Cos Z} = \cot \text{ZB} \tan \text{ZP}$$
$$\text{or Cos AZ} = \tan \text{alt} \cot \text{lat } (d)$$

(3) When at C, on the equator.

$$\text{Cos ZC} = \cos \text{ZE} \cos \text{EC}$$
$$\text{or Sin alt} = \cos \text{lat} \cos \text{hour angle } (e)$$

(4) When at D, on the horizon.

$$\text{Cos NPD} = \cot \text{PD} \tan \text{PN}$$
$$\textit{ie,}\text{— Cos DPZ} = \tan \text{dec} \tan \text{lat}$$
$$\textit{ie} \text{ Cos hour angle at sunset} = -\tan \text{dec} \tan \text{lat } (f)$$

Position of the Sun on the Ecliptic.

Fig. 48.

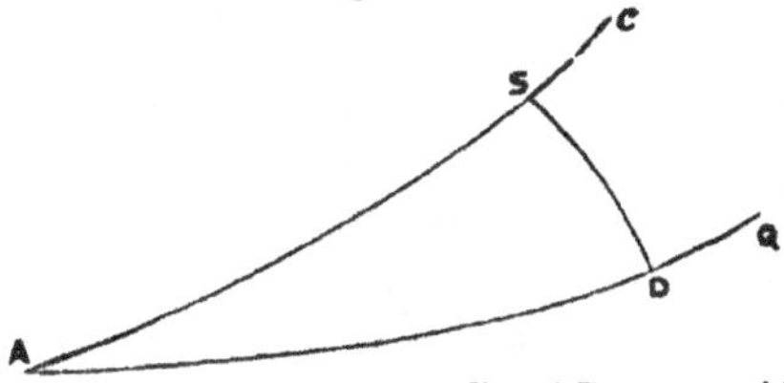

Let A*c* represent an arc of the ecliptic, AQ of the Equator, S the place of the Sun, SAD the obliquity of the Ecliptic, then AS is the Sun's longitude, SD the declination, and AD the right Ascension.

and, $$\text{Cos AS} = \cos \text{AD} . \cos \text{DS}$$
$$\text{or Cos long} = \cos \text{R}a . \cos \text{dec } (g)$$

also, $$\text{Cos SAD} = \cot \text{AS} . \tan \text{AD}$$
$$\text{or Cos } \omega = \cot \text{long} . \tan \text{R}a$$

ω being taken to represent the obliquity of the Ecliptic.

Position of any Object out of the Ecliptic.

Fig. 49.

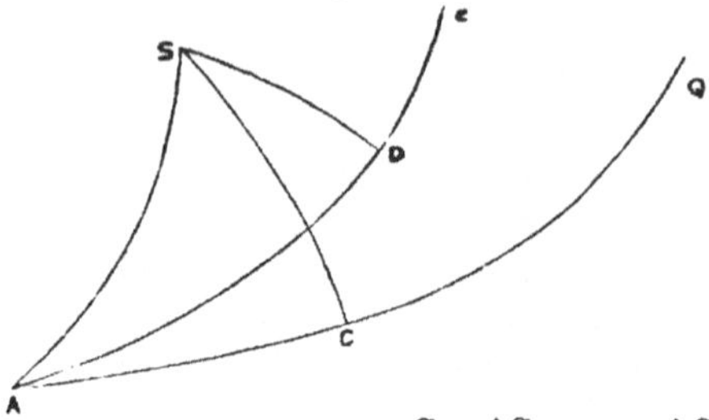

If Ae be an arc of the Ecliptic, and AQ an arc of the Equator, and S the place of any heavenly body, then SD is its latitude, AD its longitude, SC its declination, and AC its Right Ascension, and the angle eAQ is the Obliquity of the Ecliptic: then,

$$\begin{aligned} \text{Cos AS} &= \cos \text{AC} . \cos \text{CS} \\ &= \cos \text{R}a . \cos \text{dec} \\ \text{and Sin AC} &= \cot \text{SAC} \tan \text{SC} \\ &= \cot \omega \tan \text{dec} \\ \text{SAC} - \omega &= \text{SAD} \\ \text{Then Cos SAD} &= \cot \text{AS} . \tan \text{AD} \\ \text{,,} \quad &= \cot \text{AS} . \tan \text{long} \\ \text{Also Sin SD} &= \sin \text{AS} . \sin \text{SAD} \\ \text{or Sin lat} &= \sin \text{AS} . \sin \text{SAD} \end{aligned}$$

Variation by Hour Angle.

See fig. for Reduction, PS the polar dist: PZ the colat: and ZPS the hour angle being known, the angle Z may be found by Napier's Analogies, thus,

$$\text{Tan} \tfrac{1}{2} (Z + S) = \cos \tfrac{1}{2} (p - l') . \sec \tfrac{1}{2} (p + l') \cot \tfrac{1}{2} h$$

$$\text{and Tan} \tfrac{1}{2} (Z - S) = \sin \tfrac{1}{2} (p - l') . \text{cosec} \tfrac{1}{2} (p + l') \cot \tfrac{1}{2} h.$$

Now, if p is greater than l', Z is greater than S

and $\tfrac{1}{2} (Z + S) + \tfrac{1}{2} (Z - S) = \text{PZS}$, the Azimuth

But if p is less than l', S is greater than Z

then $\tfrac{1}{2} (S + Z) - \tfrac{1}{2} (S - Z) = \text{PZS}$, the Azimuth

and $180° - \text{PZS} = \text{AZS}$

which is the azimuth contrary to the latitude.

Angle of the Vertical.

Fig. 50.

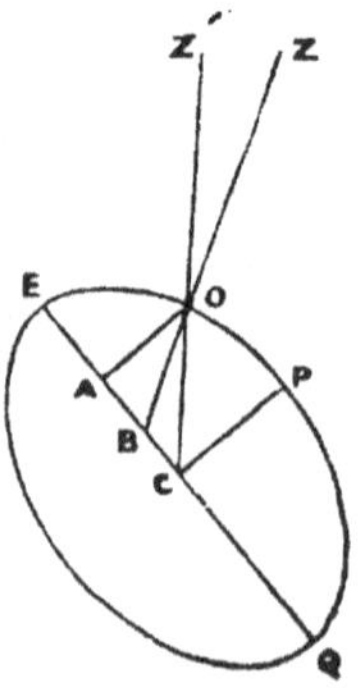

On account of the spheroidal figure of the earth, the radii are not perpendicular to the surface, *e.g.*, the line CO is not perpendicular to EOP, and therefore is not perpendicular to the sensible horizon, excepting at the equator and poles. In the figure let BO be a line at right angles to EOP; this produced will pass through the zenith of the observer at Z, while CO produced passes through Z′, named the reduced zenith. The angle ZOZ′ is called the angle of the Vertical, or "Reduction of Latitude."

Let the angle Z′CE be denoted by l', and ZBE by l

Draw OA perpendicular to EC

$$\text{Then } \frac{OA}{AB} = \tan l\text{; and } \frac{OA}{AC} = \tan l'$$

$$\text{Therefore } \frac{OA}{AB} \times \frac{AC}{OA} = \frac{\tan l}{\tan l'} \text{ or } \frac{AC}{AB} = \frac{\tan l}{\tan l'}$$

$$\text{But by a property of the ellipse, } \frac{AC}{AB} = \frac{CE^2}{CP^2} = \frac{a^2}{b^2}$$

$$\text{Hence } \frac{\tan l}{\tan l'} = \frac{a^2}{b^2}$$

$$\text{Whence } \tan l' = \frac{b^2}{a^2} \tan l$$

The fraction $\frac{b^2}{a^2}$ represents the earth's compression, computed at $\frac{1}{300}$

Augmentation of the Moon's Semi-diameter.

(See figure for Parallax in Altitude.)

The Semi-diameter of the Sun or Moon, as seen from the surface of the earth, is greater than it would be if seen from the centre.

Let the Semi-diameter as seen from C be denoted by S

and " " " O " S + a

Then OA is to AC in the ratio of S to S + a

and a is the augmentation

The angle ZOA = z the Zenith distance of A

" A = parallax (p)

Then OCA = $z - p$

$$\text{and } \frac{AC}{AO} = \frac{S + a}{S} = \frac{\text{Sin AOC}}{\text{Sin ACO}} = \frac{\text{Sin AOZ}}{\text{Sin ACO}} = \frac{\text{Sin } z}{\text{Sin }(z - p)}$$

whence,

$$S.\text{Sin }(z - p) + a \text{ Sin }(z - p) = S.\text{Sin } z$$

$$\text{and } a = \frac{S.[\text{Sin } z - \text{Sin }(z - p)]}{\text{Sin }(z - p)}$$

$$\text{or } a = \frac{2S.\text{Sin}\frac{p}{2}.\text{Cos}\left(z - \frac{p}{2}\right)}{\text{Sin }(z - p)}$$

CYCLONES.

Cyclones or Hurricanes are masses of air moving forward over certain portions of the earth, and having also a rotatory motion. Their movement is regulated by the law, that, where the barometrical pressure is high,—or, in other words, where there is a surplus of air,—out of such space winds will blow in all directions: the direction of the wind, therefore, in all cases, is towards the vacuum. The wind does not, however, blow directly from the region of high to that of lower pressure, but in the northern hemisphere, the region of lowest pressure is to the left of the direction in which it blows, and to the right in the southern hemisphere, making an angle of 70° or 80° with the circular curves which mark out the cyclone.

The direction in which the storm is advancing may be easily remembered by Buys Ballot's Law of the Winds, which is as follows ;—stand with your back to the wind, and the lowest barometer, or centre of depression, will be to your left in the northern hemisphere and to your right in the southern; or conversely, stand with the high barometer to the right, and the low to your left in the northern hemisphere, and the reverse in the southern and the wind will blow on your back.

The extent over which storms are spread is variable, being seldom less than 600 miles in diameter, and sometimes twice or thrice that amount. At the centre of this circular or elliptical whirling mass of air is a space of comparative calm, on the edge of which is the area of greatest force of wind, the force diminishing as we approach the outside of the circle, or rather as the circumference passes over the observer. When a storm contracts in area, the centre is filling up and it is dying away. The lowest barometer lies nearest the centre, the pressure increasing towards the edge : hence as the centre approaches the barometer falls, and when it is past again rises. The average rate of motion of a cyclone over Europe has been found to be 18 miles an hour; sometimes only 15, and on rare occasions 45. The direction of the rotatory motion in the northern hemisphere is opposite to that of the hands of a clock or watch, *i.e.*, from east through north to west, but the same direction as the hands of a watch, in the southern hemisphere. It is a common error to imagine, that, the fall of the barometer at any one place indicates the approach of a cyclone : this is incorrect : the occurrence of a storm is determined neither by the rapid fall nor rise of the barometer, but by the simultaneous occurrence of great differences of pressure between places not far distant from each other.

About half the storms of middle and northern Europe travel from SW or WSW, towards NE or ENE, and nineteen out of every twenty travel from some point of the compass between NE and SE. The direction of the veering is from SE through S, leaving at NW : the intensity of the gale is generally at SW, or WSW.

The storms of America generally rise in the great plain to the east of the Rocky Mountains and, advancing generally in a NE direction pass over the United States. The typhoons of the China Sea have their origin in the ocean to the east of China : they thence proceed from ENE or WSW or thereabouts : their period of occurrence is from May to October, but they are most intense and frequent between July and September.

The force of the wind is estimated by the following table known as "Beaufort's Scale."

Scale 0 to 6	Pressure in lbs. per square foot.	Velocity in miles per hour.	Beaufort's Scale.			
0·0	0·00	0·0	0	Calm	Just sufficient to make steerage way.	
0·5	0·25	7·1	1	Light Air		
1·0	1·00	14·1	2	Light Breeze	With which a ship with all sail set would go in smooth water.	1 to 2 knots
1·5	2·25	21·2	3	Gentle Breeze		3 to 4 „
2·0	4·00	28·3	4	Moderate Breeze		5 to 6 „
2·5	6·25	35·4	5	Fresh Breeze	In which she could just carry . .	Royals, &c.,
3·0	9·00	42·4	6	Strong Breeze		Single Reefs & T.G. Sails.
3·5	12·25	49·5	7	Moderate Gale		Double reefs & jib, &c.
4·0	16·00	56·6	8	Fresh Gale		Triple Reefs &c.
4·5	20·25	63·6	9	Strong Gale		Close Reefs and Courses
5·0	25·00	70·7	10	Whole Gale	In which she could just bear close-reefed Maintopsail, and reefed Foresail.	
5·5	30·25	77·8	11	Storm	Under Storm Staysails or Topsails	
6·0	36·00	81·8	12	Hurricane	Bare Poles.	

USEFUL RULES AND FORMULÆ.

A Nautical mile (the earth's diameter being 7920 miles) will be . . . . = 6080 feet.

The English Statute mile . . , . . . = 5280 feet.

Statute mile = ·8684 Nautical miles.

One pound Avoirdupois			=	7000 grains Troy
lbs. Avoir.	×	1·2157	=	lbs. Troy
ozs. „	×	·915	=	ozs. „
lbs. Troy	×	·8223	=	lbs. Avoir.
ozs. „	×	1·1	=	ozs. „
grs. „	×	·03657	=	drachms Avoir.
lbs. Avoir.	×	·00893	=	cwts.
lbs. „	×	·000447	=	tons (nearly)
Imperial galls.	×	·9834	=	Old ale galls.
„	×	1·20032	=	Old wine galls.
Old ale galls.	×	1·01704	=	Imperial galls.
Old wine galls.	×	·83311	=	„ „
Cubic feet	×	6·23210254	=	„ „
Cubic inches	×	·003607	=	„ „

227·274 cubic inches = one Imperial gallon.
A surveying chain = 22 yards of 100 links
Each link = 7·92 inches

An Imperial gallon of oil	weighs	9·32	lbs.	Avoir.
1 „ „ distilled water	„	10	„	„
1 „ „ sea water	„	10·32	„	„
1 „ „ proof spirits	„	9·3	„	„

The Area of a Circle = circumference × $\frac{1}{2}$ radius
„ „ = diam² × ·7854
= πr^2 or 3·1416 × rad²
The Area of a rectangle = length × breadth.
„ triangle = $\sqrt{S(S-a)(S-b)(S-c)}$
„ „ = $\frac{1}{2} bc \sin A$
„ „ = $\frac{1}{2}$ base × perpendicular

Area of a regular polygon = $\frac{1}{4} na^2 \cot \frac{180^\circ}{n}$
Where there are n sides each = a.

Area of an Ellipse = $\frac{\pi}{4} dd'$, where d is the major, and d' the minor diameter.

Annulus of a circle = $\frac{\pi}{4} (d + d')(d - d')$ where d and d' are the greatest and least diameters.

Surface (convex) of a cylinder = $\pi d . h$, where d = diameter of base h = height.

Surface (convex) of a cone = $\frac{\pi}{2} d . h'$, where h' is the slant height.

Surface (convex) of a sphere = πd^2 where d is the diameter.

Volume of a rectangular parallelopiped.
= a, b, c (sides)
„ „ Cylinder = $\frac{\pi}{4} d^2 h$
Where d = diameter of base, and h = height.
Volume of cone = $\frac{1}{3}$ of its circumscribing cylinder.
„ sphere = $\frac{1}{6}\pi d^3$ diameter d.
„ paraboloid = $\frac{1}{2}$ of its circumscribing cylinder.

ANGLE OF RUDDER,—SCREW PROPELLER.

In order to raise a kite to its greatest altitude, the most advantageous angle for the kite to form with the horizon is 54° 44′, which is the same as the rudder of a ship should make with the keel in order that the vessel may be turned with the greatest facility, supposing the current to have a direction parallel with the keel; and the same that the sails of a windmill, and the vanes of a smoke jack, or of a screw propeller should make with the plane of their rotation.

STRENGTH OF ROPES.

Rule.—Multiply the circumference of the rope in inches by itself, and the fifth part of the product will express the number of tons the rope will carry.

For example.—A six inch rope . . . $\frac{6^2}{5} = 7\frac{1}{5}$ the number of tons such a rope would carry.

To find the weight of a shroud laid rope.—Multiply the circumference in inches by itself: then multiply the product by the length of the rope in fathoms, and divide by 420. The product will be the weight in cwts. Thus, the weight of 120 fathoms of shroud laid rope.

$$= \frac{6^2 \times 120}{420} = \overset{\text{Cwt. qr. lbs.}}{10\ 1\ 4} = 10{\cdot}3 \text{ nearly}$$

To find the weight of cable-laid cordage.—Multiply the circumference in inches by itself and divide by 4, the result will be the weight in cwts of a cable 120 fathoms long, from which the weight of any other length may be reduced.

Thus the weight of 120 fathoms of 12 inch cable.

$$= \frac{12^2}{4} = 36 \text{ cwts} = 1 \text{ ton} . 16 \text{ cwt.}$$

Table showing the comparative strength and weight per fathom of hemp and wire rope.

Size in inches.	Hemp rope weight per fathom.	Equal to a strain of	Size in inches.	Wire rope weight per fathom.	Equal to a strain of
	lbs. ozs.	Tons. cwts.		lbs. ozs.	Tons. cwts.
3	2 4	1 16	$1\frac{1}{4}$	1 4	2 10
4	3 15	3 4	$1\frac{1}{2}$	1 9	3 10
5	6 0	5 0	$1\frac{3}{4}$	1 14	6 15
6	9 0	7 4	2	2 2	8 0
7	12 3	8 4	$2\frac{1}{4}$	2 9	8 11
8	14 3	12 16	$2\frac{1}{2}$	4 1	9 18
9	19 6	16 4	3	5 4	15 6
10	25 0	20 0	$3\frac{1}{2}$	7 1	24 6
11	30 0	24 4	4	11 6	29 5
12	36 8	28 16	$4\frac{1}{2}$	15 12	35 4

It is said that the cohesive power of a rod of English iron one inch square is equal to 55·872 lbs. Avoir.: and as the area of a square whose side is one inch is equal to the area of a circle whose diameter is equal to 1·128 inches nearly, it follows that a rod of $1\frac{128}{1000}$ is equal to the above strain, or, we may say, in practice, $1\frac{1}{8}$ in rod iron (English) 55·872 lbs. or 24·94 tons.

And, supposing the above data to be correct, the following table has been computed on the principle that the areas of circles are proportional to the squares of their diameters.

Diameter of Rod.	English Iron Strain in Tons.	Swedish Iron Strain in Tons.
$\frac{1}{4}$ in.	1·22	1.54
$\frac{1}{2}$ "	4·45	6·3
$\frac{5}{8}$ "	7·65	9·87
$\frac{3}{4}$ "	11	14
$\frac{7}{8}$ "	15	19·3
1 "	19·6	25·3
$1\frac{1}{8}$ "	24·8	32
$1\frac{1}{4}$ "	30·62	39·5
$1\frac{1}{2}$ "	44·12	56·9
$1\frac{3}{4}$ "	60·02	77·4
2 "	78·4	100·1
$2\frac{1}{4}$ "	99·23	128
$2\frac{1}{2}$ "	122·5	158
$2\frac{3}{4}$ "	148·2	191
3 "	176·4	227·5

A vertical rod of $1\frac{1}{8}$ inch English iron 16938 feet long would be rent asunder by its own weight. Also Swedish iron of 19740 feet in length, and of the same diameter.

Remarks upon a Four-fold Purchase.

It is usual, in calculating the power of purchases to divide the strain equally amongst the several parts of the fall. Such a division of strain must be incorrect, as the strain is always greatest on the running part, and, after the first friction is overcome, it appears probable that the increase of strain on the several parts will be in the ratio of their respective velocities. An experiment was made when the parts were at rest, or rather, when the first friction was just overcome. The purchase four-fold with the standing part aloft making in all nine parts,—50·5 lbs. on the fall gave 12 lbs. on the ninth part, 28 lbs on the seventh, 32·4 on the fifth, and so on; so that if we had 80 tons to raise and wished to know what sized rope would be equal to it, (when rove as a four-fold purchase):—Let C be the circumference of the rope required: then, according to the usual rule, we may say, $\frac{9C^2}{5} = 80$ tons or $C^2 = \frac{400}{9}$ and $C = \sqrt{44 \cdot 4} = 6 \cdot 6$, say $6\frac{1}{2}$ rope nearly, supposing the strain to be equally divided. But the strain on the ninth part is to that on the running part as 12 is to 50. Divide 80 in that proportion, and it gives, roughly, $6 + 14 + 16 + 19 + 25 = 80$, so that the running part must bear 25 to raise 80 tons: then $\frac{C^2}{5} = 25$ tons $\therefore C^2 = 125$ and $C = \sqrt{125}$ or $C = 11$ nearly, so that instead of a $6\frac{1}{2}$ rope, an 11 inch rope would be required.

ON CASK GAUGING.

It is usual to divide casks into four varieties,—which are judged of from the greater or less apparent curvature of their sides, viz,—

1. The middle frustrum of a spheroid.
2. The middle frustrum of a parabolic spindle.
3. The two equal frustrums of a paraboloid.
4. The two equal frustrums of a cone.

I. To find the contents of a cask of the first form.

Rule.—To the square of the head diameter add double the square of the bung diameter, and multiply the sum by the length of the cask. The product divided by 1077 = old ale gallons; or divided by 1059·1218 = imperial gallons.

II. To find the contents of a cask of the second form.

Rule.—To the square of the head diameter, add double the square of the bung diameter, and from the sum take $\frac{4}{10}$ of the square of the difference of the diameters. Then multiply the remainder by the length, and the product by ·0009 = old ale gallons, or the product by ·0009406 = imperial gallons.

III. To find the contents of a cask of the third form.

Rule.—To the square of the bung diameter, add the square of the head diameter: multiply the sum by the length, and the product again by the length, and the product again by ·0014 for old ale gallons, and by ·0014263 for imperial gallons.

IV. To find the contents of a cask of the fourth form.

Rule.—Add the square of the difference of the diameters to three times the square of their sum: then multiply the sum by the length, and the product again by ·000232 for old ale gallon, or by ·000236 for imperial gallons.

V. To find the contents of a cask from four dimensions.

Rule.—Add together the squares of the bung and head diameters, and the square of double the diameter taken in the middle between the bung and head: then multiply the sum by the length of the cask, and the product again by ·0004 for ale gallons, or by ·0004718 for imperial gallons.

VI. To find the contents of a cask from three dimensions.

Rule.—Add into one sum 39 times the square of the bung diameter, 25 times the square of the head diameter, and 26 times the product of the two diameters: then multiply the sum by the length, and the product again by $\cdot\frac{00034}{9}$ for wine gallons, or by ·0000315 for imperial gallons.

Note.—This last is the most exact rule for three dimensions only.

VII. To find the ullage of a cask standing on its head.

Rule.—Add together the square of the diameter at the surface of the liquor: the square of the diameter of the nearest end: and the square of double the diameter taken in the middle between the other two: then multiply the sum by the distance between the surface and nearest end, and tho product again by ·0004 for ale gallons, or by ·000474 for imperial gallons, in the less part of the cask, whether full or empty.

EXAMPLES FOR EXERCISE.

1. Two places A and B are 10 and 15 miles respectively from C, from which they bear N b E and NE. How far north must a vessel sail that they may appear to be in a straight line, and what is the bearing of B and A from that point?

Ans. 9·628 miles; N 84° 44′ E.

2. B bears from A, ENE 10 miles. How far South must a ship sail that B may bear from her N b E?

Ans. 42·62 miles.

3. B is a point 10 miles ENE of A, and C is a point NW of B and NNE of A. Find the distance of another point D from A B and C which bears due North of A and WNW of B.

Ans. 7·653; 10; and 4·142 miles.

4. A ship sails from A, latitude 30° 43′ N, SSE 25 miles to C where she meets a vessel which has sailed from B, SW 47 miles: find the lat of B and C.

Ans. 30° 53′ N, 30° 20′ N.

5. A fort bore from A, NE 10 miles; what would be its bearing and distance, after sailing ENE 15 miles to B.

Ans. N 78° 54′ W: 6·916 miles.

6. A tower bore from A, NE 2 miles. Find its bearing and distance from a point B, 3 miles due South of A.

Ans. N 17° 46′ E: 4.635 miles.

7. A hill bore from A, ENE, and its angle of elevation was 15°; and from a point B, due South, and distant $1\frac{1}{2}$ miles from A, it bore NE b N; find its angle of elevation at B, and also its height.

Ans. 9° 9′: 2122·032 feet.

8. Wishing to know the distance between two places on a coast, I found the first bearing NNW $\frac{1}{4}$ W, and the second NE. After sailing due east

4 miles, the first bore WNW, and the second N $\frac{1}{2}$ E; find their distance apart.

Ans. 5·91 miles.

9. The angle of elevation of a balloon from a certain point was 45° and it bore NE. From a second place 1500 yards due east of the first it bore NNE: find its angle of elevation from this position, and also its height above the sea, the height of the eye being 6 feet.

Ans. 52° 34′: 3623 yards.

10. Wishing to find the height of a monument on a hill 800 feet above the sea level, I measured a distance of 1500 feet from its base, and then found the vertical angle subtended by the monument at my position to be 2° 37′: required its height.

Ans. 90 feet.

11. Sailing along a coast, we observed the report of a gun fired from a fort bearing NE of us to follow the flash at the end of 5 seconds. After having sailed NNW for some time, the report reached us 8 seconds after the flash. Find the distance sailed, and the bearing of the fort at the second observation, sound being supposed to travel at the rate of 1142 feet per second.

Ans. 1·826 miles; S 57° 46′ E.

12. A and B are two places 3 miles apart. A's bearing from B is NNW, and the bearings of each from my position, are NNE and NE $\frac{1}{2}$ E respectively: find my distance from each.

Ans. 6·09 and 4·5 miles.

13. A Church Tower on a hill bore NE from a certain station, and its angle of elevation was 10°. 4700 feet E b S of this position, it bore NNW: find its height above the sea, and distance at each observation.

Ans. 745·8 feet: 4230 feet.

14. Wishing to ascertain the distance of three stations A, B, and C, from each other, I selected a position from which I found their respective distances to be 4250, 5000, and 2764 yards, and their bearings, NW, NNE, and N b W: find their distances apart and bearings from each other.

Ans. AB 5177: AC 2483: BC 3107 yards,

Bearing of A from B, S 71° 50′ W: A from C, N 83° 12′ W: B from C, N 52° 7′ E.

15. It is required to find the distance of four stations A,B,C and D from one another, having given the diagonal distances AD and BC, 5000, and 4164 yards, and also the following bearings, viz,—Bearing of A from C, N $\frac{1}{2}$ W: of B from C, NNE: of B from D, NNW; and of A from D, NW, D lying due East of C.

Ans. AB 1966: BD 4164: DC 3187: AC 3552 yards.

16. A mountain top is observed to bear NE b E $\frac{1}{2}$ E from a certain point, and its angle of elevation is 40°. From another station 3·5 miles ESE of the former, its summit bore N $\frac{1}{2}$ W : required its height.

Ans. 2·803 miles.

17. Having sailed NNE from A to B, I observed a point of land which bore ENE from B, whose known distance from A was 25 miles, and bearing NE. In what time could I reach it from B on the above course, at the rate of $5\frac{1}{2}$ miles per hour.

Ans. $2^h\ 27^m\ 36^s$.

18. Sailing along a coast, an object on the shore appeared to be in one with another object farther inland, bearing NE $\frac{1}{2}$ E. After proceeding ESE, 8 miles, the first bore NNW, and the second N $\frac{1}{2}$ E : find their distance apart.

Ans. 4·91 miles.

19. B,C and D are three places on my horizon. B is due north of D from which C bears NE, while C and D are in a straight line on opposite sides of my position. If C be 18 miles from B, find the distance of my boundary of view on either side, and that of D from B and C.

Ans. Boundary 25·45 miles : BC = BD = 18 miles.

20. To determine the distance of B from A and C on a coast, a boat was moored due south of A, from which position B bore NE, and C, NE b E. The boat was then moved due East 7000 yards, when the bearings of A,B and C were NW, N $\frac{1}{2}$ W, and NNE respectively : find also the bearings of A and C from B.

Ans. AB 6404 : BC 3308 yards.
Bearing of A from B N 84° 23′ W.
" C from B N 88 21 E.

21. From two stations A and B, lying east and west 1000 yards apart, the bearings of a distant object C were NNE $\frac{1}{2}$ E, and N $\frac{1}{2}$ E, while from a third station D, ENE of B, it bore NW. Find the distance of C from A, B and D, and also the distance BD.

Ans. AC 2600 ; CB 2304 ; CD 2200 ; BD 1928 yards.

22. The main road from A to B is on the arc of a circle 16 miles in diameter, and extends 15 miles. The bearings of A and B from the centre of the circle are WSW and N b W respectively. What distance will be saved by following a direct road between the two places ?

Ans. 2·63 miles.

23. B is situated 50 miles due south of A. A vessel leaves A for B, sailing SSW, then SE b E, crossing the meridian of the places at 30 miles from A, and finally reaching B on a SW course : find the distance the vessel has sailed.

Ans. 68·51 miles.

24. Two vessels sail from A and B, SE b E and SW, 40 and 60 miles respectively, meeting each other at C. Find the departure made by each, and the bearing and distance of B from A.

Ans. Departures 42·43 and 33·26 miles. Bearing of B from A, N 75° 5′ E : dist 78·30 miles.

25. A vessel steering NNE for a certain port, distant 15 miles, is carried by a current NE to a point 10 miles from her destination. Find the distance sailed, and the course she must steer to reach it.

Ans. Distance 22·04 miles : course N 80° 2′ W.

26. At 260 feet from the rock on which a lighthouse stands, its angle of elevation is double that at a distance of 300 feet farther away : required the height of the light above the sea.

Ans. 149·4 feet.

THE FOLLOWING QUESTIONS ARE CHIEFLY ANALYTICAL,—

27. When the d lat is half the departure, show that $\sin \text{co} = \frac{2}{\sqrt{5}}$

28. Given the sum of the departure, d lat and distance 120 and the difference of the squares of departure and distance 1600 miles: find the course.

Ans. $\text{Sin co} = \frac{3}{5}$

29. A vessel sails at an angle of 60° from a certain meridian which she reaches again in two boards, the second distance being double the first ; find her last course.

Ans. $\text{Sin co} = \frac{\sqrt{3}}{4}$

30. A ship sails S 45° W, b miles, then in the south-east quarter, crossing the meridian started from at distances $b\sqrt{3}$ and a miles, respectively, on either side of it : she then reaches the meridian again by sailing c miles ; show that her difference of latitude is represented by,

$$\frac{\sqrt{6}\,\overline{c^2 - a} + a\sqrt{5}}{6} + \frac{b\,(\sqrt{5} + 1)}{\sqrt{2}}$$

31. If L and L′ be the latitudes of two parallels on which meridian distances $2a$ and a are measured, show that,

$$\text{Cot}\,\frac{L + L'}{2} = 3\,\text{Tan}\,\frac{L - L'}{2}$$

32. If one meridian distance be four times another, and the difference of latitude between the parallels be 60°, show that tan lat of the lower parallel is $\frac{1}{2\sqrt{3}}$

33. If $l\ l^1\ l^2$ be the latitudes of three parallels whose meridian distances are as 1 : 2 : 4, show that.

$$\frac{\operatorname{Tan} \frac{l^1 + l}{2}}{\operatorname{Tan} \frac{l^2 + l^1}{2}} = \frac{\operatorname{Tan} \frac{l^2 - l^1}{2}}{\operatorname{Tan} \frac{l^1 - l}{2}}$$

34. Two vessels sail from opposite parallels, passing through the same middle latitude, and each reaching the other parallel. If the difference of their courses be 15°, find the courses, their differences of longitude being as 1 : 2.

Ans. 32° 58′, and 17° 58′

35. If two ships sail from latitude 40° N till their departures are as 1 : 2 and middle latitudes as 3 : 1 having made the same difference of longitude, find their latitudes arrived at.

Ans. 1° 4′, and 84° 12′.

36. Great circles from A and B cross each other in a common vertex. If a° and b° denote the distances from A and B to the vertex, and a and β the differences of longitude between A, B and the vertex, show that

$$\frac{\operatorname{Sin}(a-b)}{\operatorname{Sin}(a+b)} = \frac{\operatorname{Sin}(\beta-a)}{\operatorname{Sin}(\beta+a)}$$

37. 45° and 60° are the lowest latitudes on two great circle tracks, the differences of whose longitudes between each and their common vertex are as 3 : 1 : find the differences of longitude.

Ans. 18° 53′, and 56° 39′.

38. A vessel sails 3600 miles on a great circle from a place whose latitude is 50°, longitude 110°, to another in latitude 65°; find the longitude of the place arrived at.

Ans. 25° 38′.

39. A ship sailing towards a certain port bearing NE, is carried by a current ENE, and sails 3 miles farther than her original distance from the port, which then bears NW : find her distance at each observation.

Ans. 36·5 and 15·12 miles.

40. A point of land bears due North from a ship at A, 10 miles, but a current setting easterly carries her to B, whence the port bears NW 8 miles; find her course and distance from A to B.

Ans. N 52° 30′ E : 7·13 miles.

41. The bearings of A and B from C are NNW and NE, and the distance of A is one half that of B; show that AB = 1·863 times AC, and that B bears N 74° 45′ E from A.

42. The bearing of A from B is SW : of C from B, SSE; of C from A, SE; show that the relative distances of A, B and C from one another are as,

$$(\sqrt{2} - 1) : 1 : \tfrac{1}{2} \sqrt{2 + \sqrt{2}}.$$

43. If A and B, on the same parallel, were removed 10° farther north, their meridian distance would be $\frac{2}{3}$ of what it now is; but, if removed 10° farther south it would be $\frac{5}{4}$ of their present meridian distance; find the latitude of the parallel.

Ans. 59° 55′.

44. A point of land P bears from A, N 75° E, a miles; and from B, due South of A, N 30° E; show that the distances AB and PB are represented by $a\sqrt{2}$, and $\frac{1}{2}\, a\sqrt{2}\,(\sqrt{3} + 1)$.

45. x and x' are two points on a great circle whose latitudes are 47° 36′ and 50° 30′, situated respectively 600 miles from A and B whose latitudes and longitudes are 40° N, 12° E, and 50° N, 86° E : find the longitudes of x and x'.

Ans. 20° 58′ E, and 29° 32′ E.

46. On a certain day, when the sun's declination was 15° N, its meridian altitude was $\frac{10}{11}$ of that when its declination was 20° N : find the latitude; (zenith north).

Ans. 55°.

47. Show that the latitude of the place is the complement of half the sum of the meridian altitudes of an object when its north and south declinations are equal.

48. In north latitude, if the north declination of an object be double the south, at noon, on two opposite days, and a, a' represent its meridian altitudes on those days, then,

$$\text{Lat} = \left(90° - \frac{2a' - a}{3}\right)$$

49. If the meridian altitude of an object on a certain day, at noon, when its declination is north, be double that on another day at noon when its declination is south, show that the lesser altitude is equal to the sum of the declinations.

50. If a, a' be two different altitudes of an object when on the Prime Vertical, whose hour angles are h and h', show that,

$$\frac{\text{Sin}\,(a - a')}{\text{Sin}\,(a + a')} = \frac{\text{Sin}\,(h' - h)}{\text{Sin}\,(h' + h)}$$

51. If the Azimuths of two objects at six o'clock be denoted by Z and Z′, and their polar distances by p, p', then,

$$\frac{\text{Sin}\,(Z' - Z)}{\text{Sin}\,(Z' + Z)} = \frac{\text{Sin}\,(p' - p)}{\text{Sin}\,(p' + p)}$$

52. In latitude 50° N, how much earlier does the sun rise when his amplitude is E 20° N, than when it is E 10° N ?

Ans. $31^m\ 32^s$.

53. On a certain day when the sun's declination was 20° N, he rose an hour earlier than on another day when it was 10° N : find the latitude.

Ans. 52° 27′.

54. If h, h' a and a' be the hour angles and altitudes of two objects whose declination is 0°, show that

$$\text{Tan}\,\tfrac{1}{2}\,(h' + h).\ \tan\tfrac{1}{2}\,(h' - h) = \tan\tfrac{1}{2}\,(a - a').\ \cot\tfrac{1}{2}\,(a + a')$$

55. In latitude 40°, when the sun's declination was 20°, altitudes of 45° and 40° were taken in the morning and afternoon; find the interval between the observations.

Ans. $6^h\ 45^m\ 36^s$.

56. In latitude 45° N, when a certain object's declination was 30° N, the ex-meridian altitude was to the meridian altitude as 2 : 3; show that

$$\text{Sin}\ 3x = \frac{\sqrt{2}\,(1 \pm \sqrt{3})}{4}, \text{ where } x \text{ is the reduction.}$$

57. On a certain day, at noon, when the sun's declination was 22° 30′ S, the shadow of a perpendicular stick 4 feet high extended 2 feet farther than that of another stick 5 feet high, and placed 6 feet from the first: find the latitude of the place, (zenith north).

Ans. 53° 28′ N.

58. In latitude 60° N, when the sun's declination was 15° N and 15° S, show that the shadows of a perpendicular object at noon, are to each other as $(\sqrt{3} - 1) : (\sqrt{3} + 1)$.

59. A tower is surrounded by a circular moat. At a feet from its base, on the side of the moat nearest it, the angle of elevation was double that from the opposite side. Show that $\sqrt{b^2 - a^2}$ represents the height of the tower, b being the breadth of the moat.

60. At a distance of b feet from a point immediately under a balloon, the angle of elevation of its centre was $\beta°$, and it subtended an angle $a°$. show that the diameter of the balloon is represented by

$$\frac{2\,b.\text{Sin}\ a}{\text{Cos}\ 2\,\beta + \text{Cos}\ a}$$

61. A distance of 400 yards was measured on either side of a river, and the opposite base-line subtended angles of 30° and 45° at either end of the base: find the breadth of the river.

Ans. 481·8 yards.

62. From two positions a feet asunder, the angles of elevation of a balloon upon the same point of the compass at each station, were 30° and 45°; show that the height of the ballon above the horizontal plane is represented by $\frac{a}{\sqrt{3}-1}$.

63. Two towers whose heights are $2a$ and a above the horizontal plane, bear from a certain point N 70° E, and N 50° W, while their angles of elevation from that point are 30° and 45° respectively: show that their distance apart is represented by $a\sqrt{13+2\sqrt{3}}$.

64. From the base of a tower upon a hill-top, distances of a and $a\sqrt{3}$ feet were measured in a straight line down the slope of the hill, at which points it subtended augles of 30° and 15° respectively: show that the height of the tower $= a$.

65. Find the angle which a statue, one-third the height of a column on which it stands, will subtend at 20 feet from its base, the angle of elevation of the column being 45°.

Ans. 11° 10′.

66. Two balloons are situated, each over a town, at elevations of a and $2a$ feet: their respective bearings from my position are N 80° W and N 40° E, and their angles of elevation 30° and 45°: prove that the distance of the towns from each other is $a\sqrt{7+2\sqrt{3}}$.

67. The angle of elevation of an inclined pole was observed at a point 20 feet from it, and on the opposite side to which it inclined. Proceeding to a point perpendicularly distant from its base, and 30 feet from my former position, I found the angle of elevation of its top to be 30°, at right angles in direction to my former position. Find the length of the pole, and its inclination to the horizontal plane.

Ans. Length 31·61 feet; inclination 37° 44′.

68. Sailing into a harbour, the mouth of which was guarded by two lofty headlands, my distance from each was a and $a\sqrt{6}$ yards, and their bearings from my position were N 20° W and N 40° E respectively, the angle of elevation of each being 15°. Find the breadth of the entrance, and the height of each headland.

Ans. Breadth, $a\sqrt{7-\sqrt{6}}$: heights, $a(2-\sqrt{3})$, and $a\sqrt{2}\,(2\sqrt{3}-3)$.

69. From a point in a road, the angle of elevation of a high mountain before me was 30° : but a feet farther on towards it, the summit was just concealed by that of a lower mountain, whose angle of elevation from this point was 45°. Find their heights, and show that the height of the former was double that of the other.

Ans. Height of higher $\dfrac{a}{(\sqrt{3}-1)}$

70. From a road running due east, three distant objects A, B, C upon the same horizontal plane are visible. At a certain point, A and B appear to be in one with each other, bearing ENE. At another station, 1000 yards on the road A and C bear NW; and at a third 500 yards farther on, B and C bear WNW; find their distances apart.

Ans. AB 46·43; BC 112·1; AC 85·8 yards.

71. The angle of elevation of an object on an inclined plane, at a feet distant, was $\beta°$, while the angle of the plane was $a°$: show that the height of the object is

$$\frac{2a \,.\, \sin\beta}{\cos(2a+\beta)+\cos\beta}$$

72. The angle of elevation of a mountain bearing due east was $a°$, while b feet south-east, its bearing was north-east, and its angle of elevation $\beta°$: show that,

$$\operatorname{Sin}\beta = \sqrt{\frac{2\sin^2 a}{1+\sin^2 a}}$$

73. Two vessels start from the same meridian, each sailing due west the same number of miles per day, reaching the same longitude in four and six days respectively, the sum of their latitudes being 90°: find their latitudes.

Ans : 56° 19′, and 33° 41′.

74. The tower of a church, built upon a cliff, vanishes at an angle of 37° from the plain below, and the angle of elevation of the weather-cock, on the spire surmounting the tower is 39°. At 80 feet farther off, the angle of elevation of the tower is 30° : find the height of the spire, and that of the weather-cock above the horizontal plane.

Ans. Height of weather-cock 212·2 feet.
Height of spire 14·7

75. Coming in from sea, we observed two lightships upon the same point of the compass, bearing from us ENE. After we had sailed ESE, b miles, the one bore NNW, and the other N: find their distances from us at the second observation.

Ans. $\frac{1}{2}\, b\sqrt{2}$, and $\dfrac{b\sqrt{2}}{(\sqrt{2}+1)^{\frac{1}{2}}}$

76. A vessel steaming due north, 12 miles an hour, sights two islands, one 4 points on the starboard, and the other two points on the port bow. After an interval of an hour and a half, the former lies due east of the steamer, and the other bears WNW: find the distance and bearing of the islands from each other.

Ans. N 82° 8′ W: dist 27·26 miles.

77. From the mast-head of a vessel, 150 feet above the sea, a light was sighted upon the horizon at 8^h 30^m P.M. The light was passed at 10 P.M.: find the vessel's hourly rate of sailing.

Ans: 10 miles an hour.

78. Being at sea, I observed three distant vessels, two of which bore from my position N b E ½ E, while the other bore NE ½ N. After I had sailed NNW ½ W 2 miles, I found the first and third bearing E b N, while the other bore ENE: find the distances of the three vessels from one another.

Ans. ·404, 1·309, and 1·174 miles.

79. The angle of elevation of an upright pole, from the opposite side to its shadow, and 20 feet from its base, was 45°. Upon the corresponding day of the following year, at the same hour, the pole had been blown over by the wind in the direction of its shadow, which fell 2 feet farther than before. Find the sun's altitude, and the lengths of the shadows.

Ans. Alt 12° 10′: lengths 92·8, and 94·8 feet.

80. A balloon, c feet in diameter, subtended an angle $2a°$, while the angle of elevation of its centre was $\beta°$: show that the height of its lowest point above the horizontal plane is,

$$\frac{c}{2}\cdot \text{Cos}\,\beta\,.\sin(\beta - a)\,.\,\text{cosec}\,a.$$

Spotted something that isn't quite right?

Let us know and we'll fix it for future reprints. And as a thank you for your diligence, we'll send you a free Sothis book of your choice.

publisher@sothis.press

This page intentionally left blank

Except of course for that line above. And this one.
And whilst we have your attention, perhaps you'd
like to follow us on Instagram @sothis.press. Go on.

Made in the USA
Las Vegas, NV
02 May 2026

46267856R20142